Lean Machines for World-Class Manufacturing & Maintenance

...a definitive guide for improving equipment operability & maintainability through applied visuals and minor modifications

By:
Robert M. Williamson
Crew Chief, SWS Pit Crew
Strategic Work Systems, Inc.

©2006 by Strategic Work Systems, Inc., P.O. Box 70, Columbus, NC 28722. Printed in the United States of America. All rights reserved. No part of this publication may be reproduced in any form or by any means, electronic or mechanical, including photocopy, recording, scanning, or any information storage and retrieval system, without written permission from Strategic Work Systems, Inc., except for the inclusion of brief quotations in a review. Requests for permission or further information should be addressed to: Publications Team Leader, SWS Press, P.O. Box 70, Columbus, NC 28722.

Examples shown and described in this book represent what may be possible in a variety of plants and facilities. They do not constitute recommendations in violation of plant, facility, company, state, or federal laws and regulations. Consult company and regulatory agencies and their policies before applying visuals and/or making minor modifications as shown or described in this book. The publisher, Strategic Work Systems, Inc. and/or the Author are not responsible for misuse, accidents, damage, and/or injuries resulting from the applications of the examples shown and described in this book.

Copies of this book are available by calling SWS Press at (864) 862-0446 or (828) 894-5338, writing to SWS Press at P.O. Box 70, Columbus, NC 28722 or by visiting us online at www.swspitcrew.com.

Visit our web site at
www.swspitcrew.com

First Edition

fourth printing 2012 (revised)

Printed in the United States

ISBN 0-9778667-0-X

What others are saying about "Lean Machines…"

"This text is a worthy addition to any equipment improvement library. The visual systems are typically overlooked when it come to plant equipment. This text provides useful and practical examples of how to determine the condition of equipment without expensive technology. The years of experience Mr. Williamson has in setting up visual systems is clearly shown in this text. The hints and tips presented in this text will prove invaluable for organizations that are beginning TPM or looking to improve an existing TPM process." -- Terry Wireman, Author, "Zero Breakdown Strategies"

"The book is chock full of pictures and drawings of actual examples--visual samples of visual management--that explain the function of each idea and the role it plays in the overall system. Minor modifications to machines are also described, but always with the idea that these modifications should enhance and simplify the process by making the machine easier to clean or inspect, reduce contamination, and heighten operator awareness. This book is an essential handbook for any serious TPM practitioner." -- Mike Gardner, CME Mitsuba

"This is the best collection of ideas for making equipment easy to operate and maintain."
-- Seminar Participant

"We literally put this book to work in our plant… and what a difference it made in performance and reliability" -- TPM Manager (U.S. manufacturing)

"Brilliant book to read!" -- L.K. (England)

"I started with the book then bought the PowerPoint CD so I could share these ideas with our equipment improvement teams in the plant" – Lean Operations Coordinator

Table of Contents

Acknowledgements v
Preface vi
Chapter 1 – Introduction to Lean Machines 1
- Visual Systems for Lean Machines 2
- Minor Equipment Modifications for Lean Machines 2
- Seven steps for Making Lean Machines 3
- Catalyst for Culture Change 3
- Activities and Points to Ponder 4

Chapter 2 – Gauges 5
- How to Apply Gauge Marking Labels 10
- Activities and Points to Ponder 12

Chapter 3 – Equipment 13
- Lubrication Labels 14
- Air Filters 18
- Machine Awareness Labels 21
- Equipment Component Labeling 22
- Equipment Simplification 27
- Activities and Points to Ponder 29

Chapter 4 - Sight Glasses & Fluid Level Tubes 31
- Activities and Points to Ponder 35

Chapter 5 - Nuts and Bolts 37
- Activities and Points to Ponder 41

Chapter 6 - Machine Guards 43
- Electric Motor Inspection Windows 46
- Coupling Guards 47
- Activities and Points to Ponder 51

Chapter 7 - Locating & Communicating Problems & Opportunities 53
- Harley-Davidson's TPM Tag 56
- TPM Problem Tag Uses 57
- Purposes of the TPM Problem Tag 57
- Suggested Use of the TPM Problem Tag 58
- What about Maintenance or Repair Work Orders? 58
- Procedure for Using the Problem Tag 59
- Closing Out a TPM Problem Tag 60
- Opportunity Tags 61
- Procedure for Using Opportunity Tags 62
- Using the Equipment Action Board 63
- Preventive Maintenance Schedules 65
- Activities and Points to Ponder 66

Chapter 8 – Checklists 67
- NASCAR Checklists 67
- Lubrication Checklist & Pictorial 68
- Daily Operating or Inspection Checklists 71
- Visual Checklist 72
- Visual Machine 73
- Cleaning and Inspection Checklists 74
- Activities and Points to Ponder 77

© 2006-2012 Strategic Work Systems, Inc.

Chapter 9 - Condition Monitoring 79
- Temperature Labels 79
- Vibration Analysis Pickup Points 80
- Attaching Vibration Analysis Pickup Discs and Target Labels 81
- Directions for Use on Motors, Gearboxes and Bearings 81
- Drive Chain Tension 83
- Visual Oil Test Cards 84
- Activities and Points to Ponder 86

Chapter 10 - Stored Rotating Equipment 87
- Shaft Targets for Motors, Gearboxes, and Fans 87
  - How to Use the Shaft Target 88
- Activities and Points to Ponder 89

Chapter 11 - Parts, Tools, and Supplies Inventory Control 91
- Hand Tools... 91
- Grease Guns 92
  - Directions for Grease Gun Sleeve Use 93
- Parts and Supplies 94
- Activities and Points to Ponder 96

Chapter 12 - Equipment Flow Diagrams 97
- Activities and Points to Ponder 99

Appendix 101
- **Featured Vendors** 101
  - Air Filters 102
  - Glo-Gauges 103
  - Sight Glass 103
  - Labelmakers and software 104

- **Insights about Lean Manufacturing and Total Productive Maintenance** 106
- **Suggested Lean Manufacturing and TPM "Historical" Reading List** 107
- **SWS Products and Publications, Catalog Price List** 108

About the Author 111

COVER: An early drawing of *Watt's Steam Engine* from the mid 1700's serves as a reminder to today's *"Lean Thinkers."* The Watt's steam engine was quite simple, very functional, built by true craftsmen, and when operated responsibly became the driving force behind the *Industrial Revolution*. A simple machine launched a global *Industrial Revolution* then. Now, we are in another global *Industrial Revolution* focused on continually improving competitiveness and market changes. For machines to be competitive they must be reliable – performing as intended, first time every time. *Lean Machines for World-Class Manufacturing and Maintenance* sets the stage for a new level of reliability using world-class simplicity.

RMW

Acknowledgements

Lean Machines for World-Class Manufacturing & Maintenance is the product of many years of working as a mechanic, a teacher, and a consultant to hundreds of companies in dozens of industrial sectors. Some of the ideas are my own. However, most of the proven approaches described in this book started with many people striving to make improvements in their workplaces and on their equipment and machines. I wish to thank the thousands of people who have attended my workshops and classes. Many of you have offered insights from your own unique perspective that led to the creation and application of surprisingly simple and yet extremely effective machine visuals and minor modifications. I am especially thankful for those of you who are "out-of-the-box thinkers" and were willing to try something new.

Although I cannot list all of the companies here (due to confidentiality agreements and space), I can mention that these proven visuals and minor modifications have been successfully applied in chemical plants, refineries, tire manufacturing plants, breweries, food processing plants, dairies, off-shore and on-shore oil and gas production, pharmaceutical and bio-tech processes, automotive assembly plants, automotive parts manufacturers, bakeries, dairies, distribution warehouses, primary metals, underground and open pit mining, electric power generation plants, water and waste water utilities, building products manufacturing, PVC extrusion plants, soft drink processing and packaging, camera and film making, paper coating, textile plants, and machined/assembled products manufacturing. *This stuff really works!*

I especially thank my daughter Sharon Putman for her work, her patience, and her persistence in helping put this book together. And a special thank you to my wife Deb who worked with the roughest of rough drafts, picked things apart, and then challenged me to get this book finished. And lastly, I thank God for the ability and the opportunity to learn from everything I do and share that knowledge with so many people around the world.

Robert M. Williamson
Columbus, North Carolina

References to commercial products in this book do not constitute an endorsement.
- *NASCAR* ® is a registered trademark of National Association for Stock Car Auto Racing
- *Velcro* ® is a registered trademark of Velcro Industries B.V.
- *Lexan* ® is a registered trademark of GE Plastics, General Electric Company
- *Tyvek* ® is a registered trademark of E.I. du Pont de Nemours and Company
- *Loctite* ® is a registered trademark of Henkel Technologies
- *Excel* ® is a registered trademark of Microsoft Corporation
- *Visio* ® is a registered trademark of Microsoft Corporation
- *Post-It* ® is a registered trademark of 3M
- *GlobalMark* ® is a registered trademark of Brady Worldwide, Inc.

Preface

Lean Machines for World-Class Manufacturing & Maintenance started out as a decade-long quest for "world-class simplicity." This led to my first book, *Visual Systems for Improving Equipment Effectiveness*, in 1998. World-class simplicity resulted in common sense, but not common practice, solutions to equipment communications. Since equipment communication has evolved to software-driven logic controlling the obvious, visual communications with our equipment has become even more important. By applying and using visuals on the equipment, we are able to communicate proper operating and maintenance information visually at the point of use. Visuals also provide reminders of important specifications that were learned in training. Visuals applied to machines, equipment, and processes remove much of the guesswork often associated with operations and maintenance.

Visuals result in equipment that is significantly easier to operate, easier to maintain, and easier to inspect and troubleshoot. Visuals have also shown that equipment-specific training can be reduced by 60 to 80 percent. All of this results in eliminating human error and improving efficiency and effectiveness.

This edition launches the next level of world-class simplicity for modern manufacturing and maintenance. I have attempted to collect and explain proven methods for simplifying the interactions between people and machines. Consistent with the principles of Lean Manufacturing, I have emphasized techniques that eliminate many of the equipment-related wastes in the workplace. "Lean" truly means doing more with less of everything – eliminating waste to reduce cost. Visuals and minor modifications greatly reduce errors and improve communications about your equipment. If we can apply and use the visuals and minor modifications in the context of Lean, we will be able to reduce operating, manufacturing, and maintenance costs. This book contains hundreds of hints and tips that will improve equipment effectiveness and simplify work.

Lean Machines are:
- Easier to operate
- Easier to maintain
- Easier to inspect
- Easier to troubleshoot
- Error free
- More reliable
- Capable of reducing equipment-specific training time by 60 to 80 percent

CHAPTER 1 - Introduction to Lean Machines

Lean Machines require less of everything to operate and maintain when compared to traditional machines: less time, less effort, less money, and fewer skilled experts. Lean Machines operate more efficiently, more effectively, and with higher levels of reliability than traditional machines. Lean Machines are essential in a Lean Manufacturing plant, a Lean Processing facility, and an equipment-intensive Lean Enterprise. Traditional machines requiring traditional forms of maintenance and operator intervention can interrupt process flows, add to overall operating expense, and be a safety hazard.

Imagine reducing your equipment-specific training time by 60 to 80 percent or more, while also eliminating equipment operations and maintenance errors. Imagine making your equipment easier to operate, easier to maintain, and easier to inspect. Imagine making your equipment easier to troubleshoot and solve problems. Figure 1 shown what is possible. Sound like a pipe dream? Well, it's not a dream, but a reality practiced at a growing number of maintenance and manufacturing locations. Some equipment manufacturers are also addressing these improvements when they design and build new equipment and machinery.

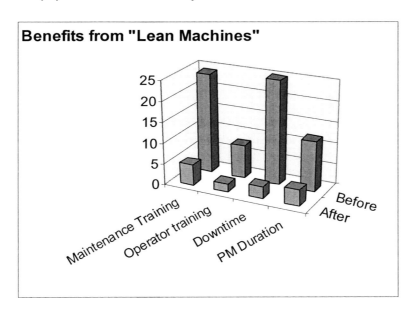

Figure 1 – What is possible with equipment visuals and minor modifications

Operability and maintainability improvement is part art and part science designed to make machines and equipment easier to operate, maintain and inspect. This is possible in part by using some simple techniques known as visual systems, which communicate specific information quickly at or near the equipment. Improved accesses to critical parts or information through minor modifications contribute the rest of the formula for successful Lean Machines.

Another gauge example…

A custom adhesive vinyl label produced on the Brady *GlobalMark* ® label maker shown in Figure 11 allows both the "normal" and the "danger" or out-of-standard ranges to be printed and cut to size and shape. *Markware*™ v 3.7 software is required.

Hint: Be sure to clean the surface of the gauge with alcohol and dry before attaching the label.

Figure 11 – Custom adhesive vinyl label (Copyright Permission Granted by Brady Worldwide, Inc.)

Go-no go gages: While we are addressing gauges, there are other types of "gages" used to detect go or no-go conditions. Figure 12 shows a NASCAR inspection that measures the rear deck lid height from the ground, one of many critical pre-race and post-race inspections.

Figure 12 – NASCAR height gage

A red band appears on the plunger at the top of the block if the height is too high or too low. As long as only the green on the plunger is visible, the deck lid height is within specifications. This is a technique that can be used on quick changeover and equipment setup.

Chapter 2 Activities and points to ponder…

1. What do the gauges in our plant or facility measure?

2. Marking and labeling gauges makes them easier to read and prevents errors. Would this be helpful in your plant or facility? Why?

3. How can gauge marking and labeling make training more efficient and effective?

4. Which gauge marking techniques would work best on the equipment in your plant or facility?
   - ☐ Description of what the gauge is measuring
   - ☐ Green operating ranges
   - ☐ Red danger ranges
   - ☐ Yellow caution ranges
   - ☐ Adjustable Gauge Marking Labels and ranges (changeable)
   - ☐ Rotated gauges with indicator needles pointing as 12 o'clock positions
   - ☐ _____
   - ☐ _____
   - ☐ _____

5. What would you do to improve these gauges in this panel? List your ideas:

   _____

   _____

   _____

   _____

Products described in this chapter available from www.swspitcrew.com:

016-1 Red Gauge Marking Label (9" x 10" sheets)
016-2 Green Gauge Marking Label (9" x 10" sheets)
016-3 Yellow Gauge Marking Label (9" x 10" sheets)

CHAPTER 3 - Equipment

Labeling and modifying equipment to make it easier to inspect, maintain, and operate includes a variety of methods. Many machines have covers, panels, and guards to protect parts of the machine, to provide safety barriers, or to improve appearances. These covers often make inspection and maintenance difficult and time consuming, frequently requiring that the equipment be shut down and locked out.

Labels

Labeling each piece of equipment or machinery with its official identification number is important. As part of the managed maintenance process, or computerized maintenance management system (CMMS), equipment history becomes more accurate, and maintenance backlogs are less cluttered if everybody calls the equipment by the correct name and I.D. number when filling out a work request.

Figure 13 shows a 4-inch label with the name and I.D. number applied:
"***#4 BRIDLE, EQUIP #3250.***"

The label is an outdoor bumper-sticker type of adhesive material made on a Brady "Globalmark" machine. (See Appendix for more information.)

Figure 13 – Labeled machine

Many electrical cabinets and disconnects in today's plants are labeled with detailed information from the schematic diagrams or piping and instrumentation diagrams. In many cases, these labels do not readily indicate the equipment that is actually being controlled. Attach labels to these electrical panels stating in everyday terms what is being controlled as shown in Figure 14.

Figure 14 – Labeled electrical disconnects

Notice how the functions of both electrical panels are very clear: "Water Pump #2 Motor Disconnect." Most people would expect Pump #2 disconnect to be on the right. However, due to the installation method in this plant, the locations are reversed.

Lubrication Labels

Labeling lubrication requirements and locations is very important. In some plants, lubrication-related failures account for nearly 70 percent of equipment downtime. Proper lubrication is made easier and downtime is reduced if the instructions are on the machines and all lube points are labeled and color coded. Figure 15 shows colored vinyl adhesive lube labels. These labels are placed next to each lube point and on lube cans and grease guns. The colors correspond to a color-coded lube standards and machine diagrams

Figure 15 - Colored adhesive vinyl lube point labels

An adhesive lube point label that is applied to a clean surface around the Zerk (grease) fitting is shown in Figure 16. This label specifies the lube type (EP2) and the lube point number that corresponds to the lube diagram and PM instructions (10). These custom-made ring labels can also be color-coded to indicate the type of grease.

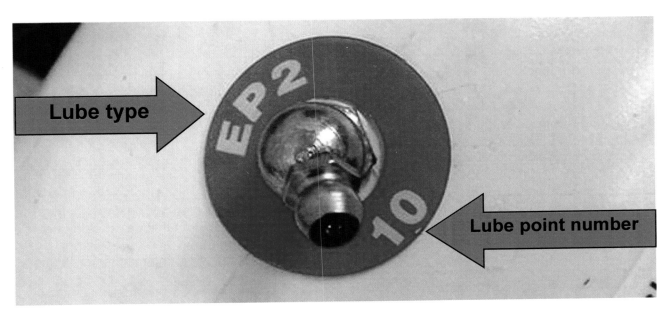

Figure 16 – Zerk fitting label

Protecting and labeling lube points can generate big savings, as well as serve as a form of error-proofing. Grease fittings (Zerk fittings) are often exposed to dirt, moisture, and other contaminants that can cause lube failure. By "ganging" grease points together, the lube tasks can also be sped up by not having to crawl all over the machine or accessing fittings in remote areas of the machine. Figure 17 shows a number of improvements for lubrication.

Figure 17 – Ganged lube fittings, protective caps, and labels

In this application, we have used the following methods:

- Zerk fittings protected with color-coded grease fitting caps
- Lube points numbered, indicating sequence or location from PM or lube diagram
- Label containing lube point number, frequency, volume of grease, and type of grease

> Hint: When "ganging" grease fittings, make sure there is a PM inspection of the lube lines that run from the fittings to the lube points (bearings, etc.). We have seen these lines broken, pinched closed, or disconnected, leading to grease piling up inside the machine.

Lubricating enclosed bearings in pumps and motors can cause serious problems and failures if not done properly. In many of these cases, there is a "relief" or vent plug associated with each grease fitting. The purpose of the relief plug is to allow old grease to run out of the bearing cavity without damaging the seals that protect the bearings and hold the grease in. Figure 18 shows a vertical turbine pump and a motor with relief plugs (previously unknown to the mechanics in this plant).

Figure 18 – Turbine pump grease relief plug and label

Many electric motors have enclosed bearings requiring removal of the "relief" plug when greasing. Locations of the relief plug may vary but are generally opposite the grease fitting. The following Figure 19 shows an example of an electric drive motor grease fitting and relief plug location.

The greasing process should be defined in the preventive maintenance for this motor and a label applied near the relief plug as a reminder.

Figure 19 – Drive motor grease fitting and relief plug

Label the grease relief plug. In both cases, the relief plug was labeled with greasing instructions as shown in Figure 20:

> **Grease Relief Plug.
> Remove when greasing. Allow old grease to run out. Run equipment until grease runs out. Reinsert and tighten plug. Clean up.**

Figure 20 – Grease relief plug label

In these cases, it is proper to grease the fitting when the equipment is not rotating or when it is running. In either case, the relief plug should be removed to allow the old grease to run out until new grease is observed. While running the motor, allow the new grease to finish running out for a few minutes and then replace the relief plug. Bearings will last longer, and motors or pumps will not be damaged because of blown grease seals.

The preventive maintenance (PM) work instructions must also be revised to include the new information on proper greasing in addition to labeling the grease relief plug with locations and instructions.

Figure 21 – Motor lube label

The lube instructions in Figure 19 are engraved on a tag, which is then glued to the motor. The red painted lines (using a paint pen) run from the label to each of the two lube fittings. The label becomes a reminder of what is on the annual PM for that motor.

Air Filters

Air filters are used for many applications to keep dirt, dust and fumes out of cabinets, computers, and motors. Ease of inspection, ease of removal and replacement is critical. Filter replacement is made faster by labeling the actual dimensions of the filter media to be cut from bulk stock, or the standard filter. Adhesive Velcro® hook tape can be applied around the filter opening to hold the filter media in place as shown in Figure 22.

Figure 22 – Filter attachment with adhesive Velcro® tape

In some cases, this will require a simple modification of the air intake as shown in Figure 23.

Figure 23 – Air intake modification for Velco® attached filter media

Labeling the sizes of the filter media makes it easier to prepare for fast filter changes by having the correct filters on hand. Figure 24 shows filter size labeling.

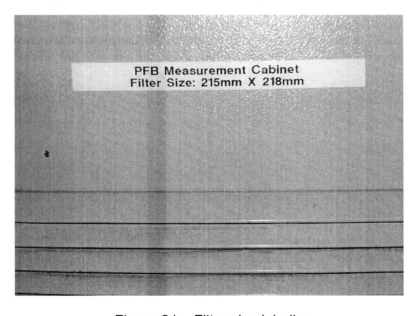

Figure 24 – Filter size labeling

Filter changes can be done based on time or condition. In most cases the filter condition is the best indicator. In Figure 25, the team labeled a newly installed motor filter with the date it was installed. Their goal was to answer the question: "*How long do these filters really last, and what contaminates them so fast?*"

Figure 25 – Filter installation date label

The motor filter in Figure 25 above is held in place with adhesive Velcro® tape around the entire circumference of both ends of the filter housing. This held the filter in place and provided a "seal" around the housing that eliminated gaps and sags in the filter media. The red tape provided an extra holding constraint as well as a place to label the date. A filter size label is on the white label shown at the upper left of the filter housing.

Machine Awareness Labels

"Just what is the cost of equipment downtime?" This question is often asked. In Figure 26, this company applied labels to their critical process equipment indicating the downtime cost. This helps people in the plant appreciate the reliability improvement methods and helps gain their buy-in. In some processes, the cost of downtime can be $50,000 or $300,000 per hour! The problem is most people who work on and around the equipment are unaware of the true cost of downtime.

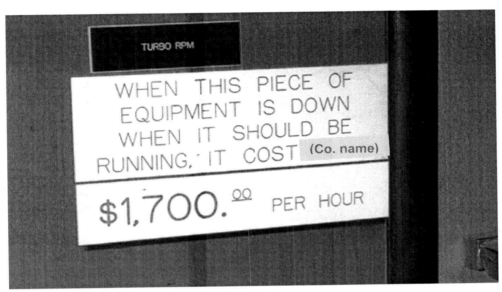

Figure 26 – Label showing cost of downtime

The cost of removing and replacing equipment components can add up. Figure 27 shows a label applied to a component that was routinely damaged by errors and abuse. Additional cost would include lost production revenue while the parts were being replaced.

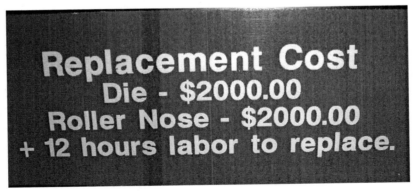

Figure 27 – Equipment component replacement cost label

Equipment Component Labeling

Another NASCAR Nextel Cup example: Labeling critical settings on the valve cover of the engine is shown in Figure 28.

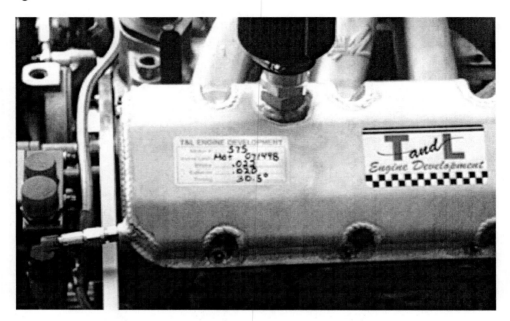

Figure 28 – NASCAR engine

The label tells the intake and exhaust valve clearance and the timing setting for this installed race-ready engine. This technique puts the critical information right where it is needed and serves as a reminder, eliminating guess work, trial and error, or going somewhere else to find the needed information.

Take the guess out of making equipment adjustments by labeling the direction that reduces (-) or increases (+) a setting as shown in Figure 29.

Figure 29 - Adjustment indicators (Copyright Permission Granted by Brady Worldwide, Inc.)

Labels on equipment components eliminate many errors and mysteries in the workplace. Electrical box labeling and valve positions are shown in Figure 30.

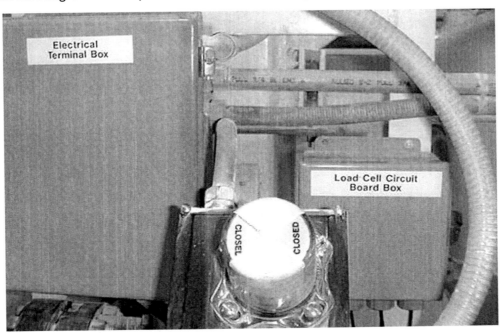

Figure 30 - Electrical box labeling and valve positions

Labeling inside machines can communicate component names and schematic and drawing file numbers. Figure 31 shows examples of these types of labels.

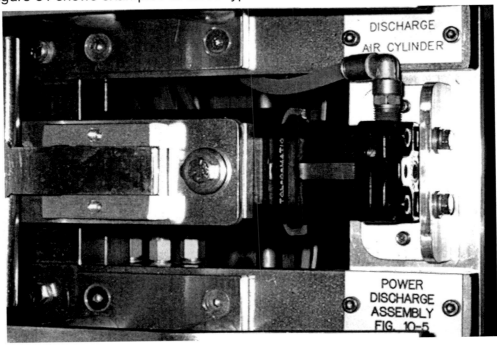

Figure 31 - Labeling inside machines

Labeling of replacement parts such as equipment/system filters are critical to proper equipment operations. There are times when substitutions are made because the correct part number is not readily available. In some of these cases, equipment is damaged. In Figure 32, the filter housing is labeled with very specific replacement filter information.

Figure 32 – Filter housing with replacement filter information

In this application, we have used the following methods:
- Filter replacement part number and original equipment number
- Price of replacement filter (to increase awareness of maintenance costs)

Piping and hoses are frequently routed from one part of the process to another in ways that make troubleshooting very difficult. In Figures 33 and 34, compressed air hoses and piping were labeled and numbered to match the valve numbers that controlled the air flow.

Figure 33 – Compressed air hose labeling

Figure 34 – Compressed pipe labeling

In these compressed air hose and piping applications, we have used the following methods:

- Each hose is labeled to match the controlling valve numbers.
- Hoses are labeled with "compressed air" and direction of air flow using a paint pen.
- Pipe is labeled with "compressed air" and direction of flow arrow using Brady labels.

Valve position in the same compressed air system was also labeled to show the normal position of the valve (Figure 35).

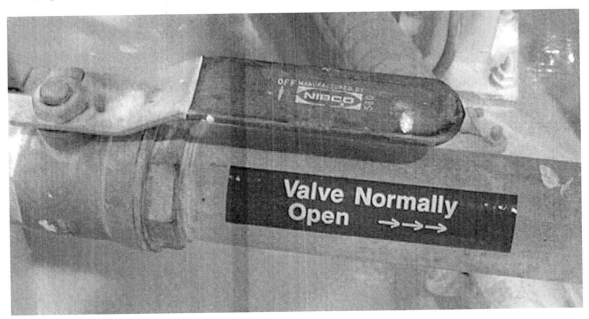

Figure 35 – Compressed air ball valve

In this application, we have used the following methods:

- Label applied directly under the ball valve handle indicating "valve normally open" with arrows pointing the direction of the handle when open

> **Hint**: When applying adhesive-backed labels, always clean the mounting surface thoroughly with solvent to remove grease, dirt, and other contaminants. This will help the labels stay in place for years.

Are those computer and PLC cabinet cooling fans really working? Small "flags" can indicate when fans are moving the air and the direction of flow as shown in Figure 36.

Figure 36 – Computer PLC Cooling Fan Indicators

Many cases of disabled fans, broken fans, and fans running in the wrong direction have been detected with these "vent fan flags" (cooling fan indicators).

Adjustments can shift due to vibrations and tension. Chain and belt drives tensioned by moving the motor and tightening the motor base maintain proper tension. Any change can be visually detected by applying a "match marking tape" after tensioning. After application, the match-marking tape must be cut where the motor base meets the base plate. Figure 37 shows an example of adjustable component match-marking tape.

Figure 37 - Adjustable component match-marking

Equipment Simplification

Much of the equipment in manufacturing plants is overly complicated for the desired function. This complexity often leads to problems, failures, continual adjustment, errors, and difficulty accessing other components. All of this contributes to higher equipment downtime and increased maintenance and operating costs.

The guillotine cutter shown in Figure 38 has one air cylinder actuating the blade through a complicated pivot/cam joint mechanism. The cylinder is mounted 90° to the cutting direction.

Figure 38 - Complex cutter mechanism

Work order history and production downtime reports indicated high downtime of this production line because of problems with this guillotine cutter.

This application of match marking on the log flaker in Figure 48 is important because access to check the hold-down bolts with a torque wrench is limited to equipment downtime only. Prior to match-marking, the machine had to be locked and tagged out and the safety gates opened for the mechanics to enter the area. With the match marks on properly tightened and torqued bolts, operators and maintenance technicians can now perform visual inspections from outside the safety rails to see if the bolts have vibrated loose. When the bolts are re-tightened and torqued, the old match mark is scraped off and re-painted.

In this application, we used the following methods:

- Proper bolt tightening and torquing to specification
- Match marks on the bolts and the pillow block housing using a paint pen

Small nuts and bolts cannot be easily marked with a paint pen. An alternative to match marking is the use of Torque Seal, as shown on the #10 hex socket head cap screws in Figure 49.

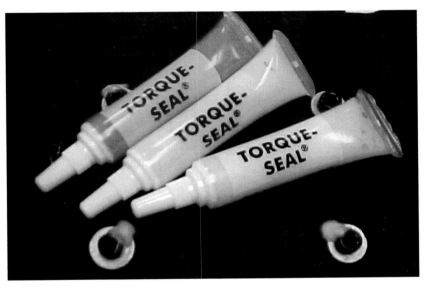

Figure 49 – Torque Seal application

Torque Seal is also known as an "anti-sabotage lacquer" used to hold positions of adjustable screws in electronics or to prevent un-detected adjustment of small screws. As an alternative to match marking, Torque Seal, available in six colors, is applied as follows on small screws and nuts used to hold small devices such as proximity switches, limit switches, and sensors in position:

- Tighten and torque the screws
- Apply a drop of Torque Seal to one side of the screw and onto the machine. After this lacquer hardens, it will crack when the screw or nut loosens. It is not designed to provide any holding power or prevent loosening.

Hint: Nuts and bolts very rarely vibrate tight. They almost always vibrate loose.

NASCAR Nextel Cup race teams continue to give us examples we can use in our plants and facilities. Figure 50 shows a simple "error proofing" technique they use with nuts and bolts.

Figure 50– NASCAR nuts and bolt error proofing

Duct tape? The race team mechanic was working on the fuel cell vent/overflow tube. He removed the nuts, bolts, and washers and stuck them to a strip of "200 mile per hour duct tape" placed on the edge of the trunk opening. These fasteners must be re-installed before the car goes to the track or a severe NASCAR penalty may be levied. It is a safety issue – a loose vent tube can fall down inside the car and on top of the fuel cell causing fumes and fire (a bomb!) inside the car.

The "error proofing strategy" here is simple. The screws, nuts, and washers on the duct tape were placed where they could be seen. If the rear deck (trunk) lid was closed, they still would be seen when the mechanic installed the hold-down lock pins.

How often in our plants and facilities are fasteners and small parts left over after maintenance? How often are small nuts and bolts misplaced? The NASCAR mechanics have found many ways to keep these small fasteners close to where they are needed, not in their pockets, not on the floor, and not in the toolboxes.

> **Hint:** Did you know that the word "racecar" is a palindrome? A ***palindrome*** is a word that is spelled the same way forwards or backwards. Can you think of others?

Chapter 5 Activities and points to ponder...

1. What are some of the problems caused by nuts and bolts on your targeted critical equipment?

2. Review your targeted critical equipment. What are the observed conditions of the nuts and bolts?
    a. _____ Loose nuts and/or bolts
    b. _____ Missing nuts and/or bolts
    c. _____ Wrong nuts and/or bolts
    d. _____ Damaged threads
    e. _____ Damaged hex heads, rounded corners
    f. _____ Damaged hex sockets or screwdriver slots
    g. _____ Painted threads making removal difficult
    h. _____ Hard to get wrenches on nuts or bolts to remove
    i. _____ (other)

2. Are torque wrenches available to mechanics? Are they used often?

Products described in this chapter available from www.swspitcrew.com:

004 Paint Pens (white, yellow, red, black, blue, green, orange)
026 Torque Seal (red, orange, lemon, green, yellow-orange, pink)

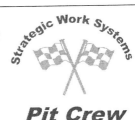

© 2006-2012 Strategic Work Systems, Inc.

CHAPTER 6 - Machine Guards

Safety guards cover parts of machines that rotate, gears that drive, belts and chains that transmit power, moving materials, and levers or pinch points. These guards are important for workplace safety, but they complicate maintenance and inspection tasks. Some guards are easily removed with minimal tools. Other guards are extremely difficult to remove because of their attachment points or other parts that are in the way.

Many bearing, chain, belt, and seal failures go unnoticed because of these well-intended guards.

Making these guards into windows is one way to continue the safety function and improve maintenance and inspection tasks. This section offers some ways to make guards more user-friendly.

The carding machine shown in Figure 51 is an example of a large guard covering a very critical series of belts, sheaves, and gears. When these belts wear and slip, the quality of the output product is damaged. In their present condition, removal of the 6-foot long guard is required to inspect the condition of the belts. A simple visual modification can make this task easier and performable by anyone who knows the equipment.

Figure 51 – Carding machine guard

CHAPTER 7 - Locating & Communicating Problems & Opportunities

When machines have problems, it is up to people to address them. Quite often, especially in a multi-shift plant or facility, the same problems get reported many times as "maintenance requests." This clutters up the maintenance management process and may lead to several work orders being issued for the same problem.

The use of problem tags and opportunity tags can eliminate the duplication of maintenance requests and provide an effective means of communications. Figure 64 show problem tags used to locate and communicate equipment problems or opportunities in a manufacturing plant in Japan.

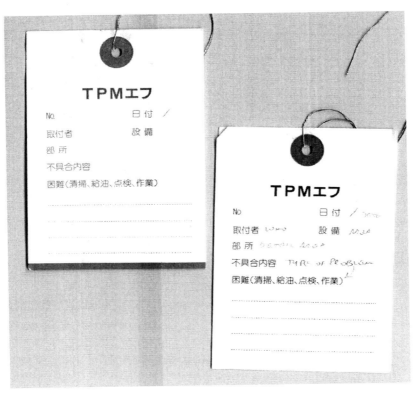

Figure 64 - Japanese TPM tags

The tag on the left in Figure 60 is printed in red ink indicating that the problem must be addressed by a maintenance worker. The black-ink tag on the right indicates that the problem can be addressed by either a maintenance worker or an operator.

Some problem tags are two-part, perforated top and bottom. The same serial number appears on both parts of the Coolant Leak and Machine Defect problem tags shown in Figure 65.

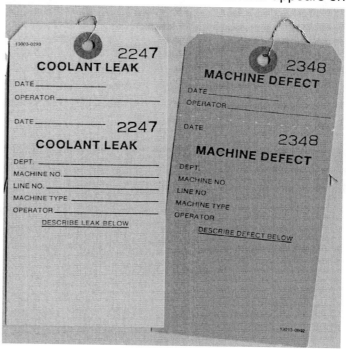

Figure 65 – TPM tag and two-part problem tags

In this example, different tag colors are used to indicate different types of problems: Coolant leak (green), Machine Defect (pink), Electrical Problem (blue), and Lubrication Problem (yellow).

Some problem tags use carbon paper between the two parts. While the back card that hangs on the machine is bright orange, the front paper is plain white as shown in Figure 66.

Figure 66 - Two-part carbon paper problem tag

Some companies have tried carbonless paper for their two-part tags as shown in Figure 67. This saves time in filling out the same information on a top and bottom portion of a two-part tag. However, the problem is that the written image on the tag left hanging on the machine deteriorates and becomes unreadable in about a week, depending on the heat and humidity.

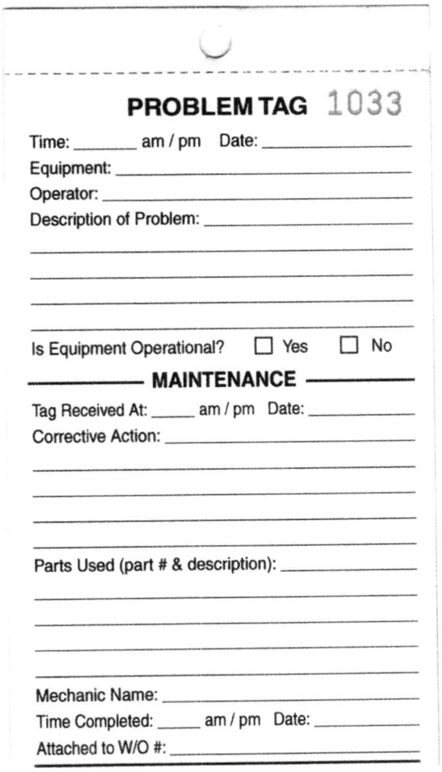

Figure 67 - Carbonless paper three-part Problem Tag

Another "Problem Tag" approach is to use two separate tags: One for identifying and communicating the problem and the second one for identifying the corrective action, or completion information. Examples are shown in Figure 68.

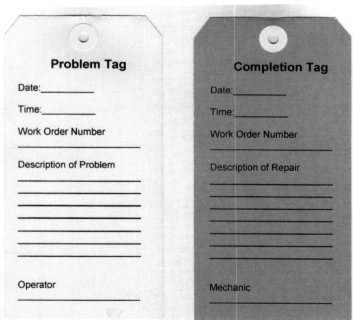

Figure 68 - Problem and Completion tags

© 2006-2012 Strategic Work Systems, Inc.

Problem tags are typically printed on a bright-colored cardstock. Our tags come in two formats: A **Problem Tag** for identifying and communicating problems, and an **Opportunity Tag** for identifying and communicating opportunities for improvements or ideas across multi-shift operations. Figure 69 shows both the two-part perforated Opportunity Tag and Problem Tag.

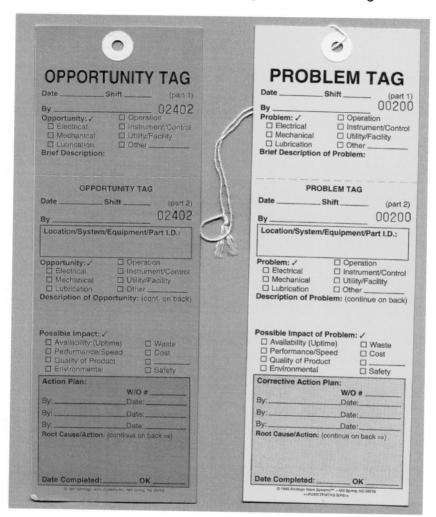

Figure 69 – Strategic Work Systems'
two-part perforated Problem and Opportunity Tags

Harley-Davidson's TPM Tag

Another type of problem identification tag was developed by Harley-Davidson for use in their engine plant in the early 1990s. This multiple-part work request tag is shown in Figure 70. This neon-green paper tag is a carbonless form. The top copy is turned in to the area supervisor or maintenance to have a work order generated. The upper right side is filled out with the brass tag number or equipment tag number, the machine name, the name of the person filling out the tag, the date the tag is filled out, and a description of the problem and its location on the machine. This information also appears on the duplicate bottom copy, which is made of heavy-stock paper. This bottom copy is hung near the machine maintenance log on the machine.

The comments portion on the left side of the tag is filled in with a brief description of the problem, torn off along a perforated edge, and hung as close to the problem as possible.

The tradesperson writes comments in the section labeled Trades Person's Comments, and the Harley-Davidson logo in the lower right corner is torn off along a perforated edge when the repair is complete.

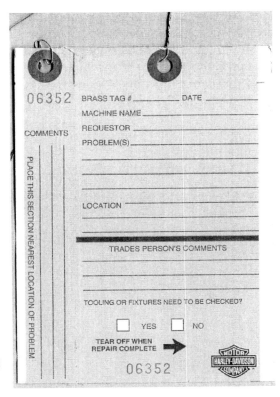

Figure 70 – Harley-Davidson's TPM tag

TPM Problem Tag Uses

The problem tag is widely used in TPM work cultures from the early stages of development to the more mature forms of TPM. It provides an easily visible method for calling attention to problems on the equipment and for communicating the nature of the problem.

When used correctly, the problem tag also becomes an excellent learning tool to help people look for and communicate problems with their equipment in ways not often practiced in traditional manufacturing and maintenance.

Purposes of the TPM Problem Tag

- It is a visible indication of where the problem is located on the equipment.
- It provides a brief description of what the problem appears to be.
- It aids in communications among all members of the work group.
- It provides documentation of all problems and improvement opportunities.

Using the Problem Tag requires that the people in the work group all understand their roles in the new TPM work culture. Initially, the tag is used when training the cross-functional work group about cleaning and inspecting equipment. In this learning activity, maintenance, operations, engineering/technical, and others work together to begin understanding the equipment and sharing information about how it is supposed to look and what causes problems.

CHAPTER 8 - Checklists

Paperwork is often seen as a necessary evil in today's workplace. Yet without proper procedures and checklists, errors are made or information is lost. This section focuses on making checklists and procedures more user-friendly.

NASCAR Checklists

Let's start with NASCAR Nextel Cup examples of checklists. These teams use checklists for every critical task pertaining to the car, the parts room, the transport truck, the shop, the pit carts, the chassis… Everything that will have an effect on the performance of the race car or the team has been reduced to a checklist of sorts.

So why this focus on paperwork? Consistency. Making sure that regardless of who does a task, or who builds a chassis, everything is done the same way and documented the same way. In industrial terms, this is often referred to as "standardized work" or "best practices."

Figure 78 shows page one of a three-page Pre-Race Check Sheet. Notice how each person has a list of things to do to get the car ready for the starting lineup? And look closely at the level of detail of each item – they are the "simple" things that experience has taught the team can and will affect the performance of the racecar.

```
                HENDRICK MOTORSPORTS
                    DUPONT TEAM
                PRE RACE CHECK SHEET 1 OF 3

CAR_____    TRACK_____    DATE_____

_____CHAD_____:                _____EDDIE_____:
____INSPECT OVERFLOW HOSE         ____INSPECT FRONT SHOCKS & MOUNTS
____GAS LINE CONNECTIONS IN TRUNK ____LOCTITE FRONT SHOCK BOLTS
____INSPECT FUEL FILTER           ____INSPECT FRONT HUB CAPS
____VENT TUBE FLAPPER AND LINE    ____TIE ROD ENDS TIGHT & SAFETY WIRED
____CATCH CAN FIT ON VENT TUBE    ____TIE ROD SLEEVES TIGHT
____HOSE CLAMPS TIGHT             ____PITMAN ARM NUT SAFETY WIRED
____FUEL CELL SLEEVE              ____CENTER LINK DOUBLE NUTTED &
____FUEL CELL HOLDDOWN BOLTS          BEARING ADJUSTMENT
____FUEL CELL TOP PLATE BOLTS     ____IDLER ARM MOUNTING BOLTS
____PRESSURE LUBE DRY BRAKE       ____STEERING BOX MOUNTING BOLTS
____DECK LID HINGES DOUBLE NUTTED ____LOCTITE STEERING SHAFT JOINTS
____DECK LID PINS, LANYARDS, PROP ____NO PLAY IN STEERING U-JOINTS
____LOCTITE REAR SPOILER AND MOUNTING ____POWER STEERING FLUID LEVEL
____NEON TAPE OVER REAR JACKSCREWS ____BALL JOINTS TIGHT & PINNED
____HOOD HINGES DOUBLE NUTTED     ____LOWER CONTROL ARM MOUNTING
____HOOD PINS, LANYARDS, AND PROP     BOLTS
____FRONT FENDER BRACES           ____GREASE FRONT END
____REAR QUARTER FENDER BRACES    ____INSPECT SWAY BAR ARMS & HELMS
____CHECK ALL SIDE SKIRTS         ____REPLACE SWAY BAR CHAIN
____JACK SCREWS & SPRING BUCKETS/CAGES ____INSPECT FRONT HUBS FOR LEAKS
____CHECK DUCTWORK, FIREWALLS ETC. FOR SEAL    OR LOOSENESS
____CHECK WINDSHIELD BOLTS FOR TIGHTNESS ____UPPER A-ARM DOUBLE NUTTED
____CLEAN & RAIN X WINDSHIELD     ____CAMBER SHIMS SAFETY WIRED
____FILE FRONT HUB CAP SCREWS     ____INSPECT HEADERS & BOLTS FOR
____CHASE WHEEL STUDS WITH DIE        TIGHTNESS
____ANTI-SIEZE STUDS              ____ENGINE FAN MOUNTING
____GLUE WHEEL SPACERS            ____RADIATOR HOSE CLAMPS TIGHT
____INSTALL NEW LUGNUTS           ____SURGE TANK MOUNTING
____MARK STUDS AND HUBS           ____SURGE TANK LINE TIGHT
____JACK POST CLEARANCE           ____OVERFLOW HOSE CLAMPED & WIRED
                                  ____PRESSURE CHECK CAP
                                  ____RADIATOR MOUNTING
```

Figure 78 – Pre-Race Check Sheet

These Pre-Race Check Sheets are then taped to the racecar for the team members to follow. After all tasks have been completed and signed off on, these sheets become part of the equipment history file for the race. They will be used again to do root cause analysis upon completion of the race.

What we have learned from the race teams is that according to Pete Bingle of Hendrick Motorsports, "People are human and we tend to forget things. These checklists are there. We use them all the time, week after week, to make sure all of the right things get done." Here are a few industrial parallels for this type of reminder.

Lubrication Checklist & Pictorial

Figure 79 shows a Machine Pictorial from the Harley-Davidson Engine Plant. These pictorials list all of the lube points and are laminated and attached to each machine. This was part of Harley-Davidson's quest for Total Productive Maintenance (TPM) and the elimination of lubrication problems on their equipment.

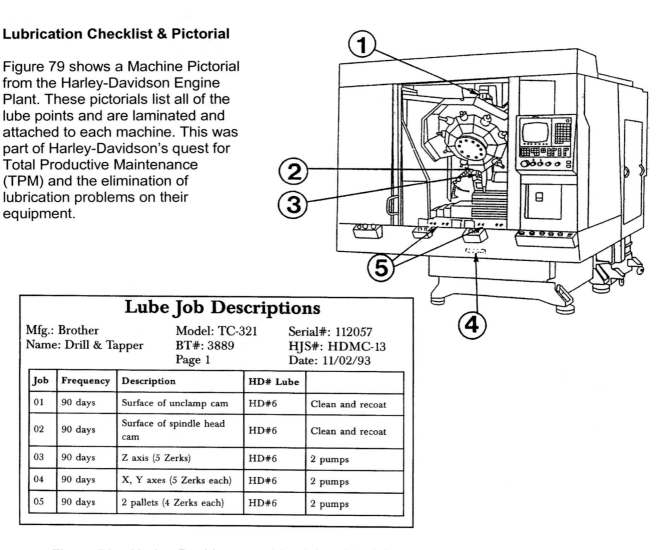

Figure 79 – Harley-Davidson machine lube pictorial

Every lube point is shown and numbered. The lube frequency, description of the points, lube type, and lube method is compiled into this simple chart.

Figure 80 shows the TPM Operator Oiling Checklist corresponding to the machine lube pictorial in Figure 78 typically posted on the **Equipment Action Board** near the machine. It is used by operators to keep track of when each lube job was performed. This helps prevent over-lubrication and maintain the equipment lubrication history. When each activity is completed for the year, the checklist is placed in the equipment file.

TPM Operator Oiling Checklist

Machine: Brother Drill and Tapper   Month:_____
Brass Tag: 3889   Dept. 904   Cell_____
Enter date and shift in box (e.g. 18/2 for 2nd shift on the 18th).

Job 1: Clean and recoat surface of unclamp cam with HD#6. _____   Jan. ☐ April ☐ July ☐ Oct. ☐

Job 2: Clean and recoat surface of spindle head cam with HD#6. _____   Jan. ☐ April ☐ July ☐ Oct. ☐

Job 3: Grease Z axis (5 Zerks) with 2 pumps of HD#6. _____   Jan. ☐ April ☐ July ☐ Oct. ☐

Job 4: Grease X, Y axis (5 Zerks each) with 2 pumps of HD#6. _____   Jan. ☐ April ☐ July ☐ Oct. ☐

Job 5: Grease 2-Pallets (4 Zerks each) with 2 pumps of HD#6. _____   Jan. ☐ April ☐ July ☐ Oct. ☐

Machine cleaning: Follow cleaning norms and initial each box after job is complete.

1st Shift								2nd Shift								3rd Shift							
Week	S	M	T	W	TH	F	S	Week	S	M	T	W	TH	F	S	Week	S	M	T	W	TH	F	S
1								1								1							
2								2								2							
3								3								3							
4								4								4							
5								5								5							

Figure 80– Harley-Davidson Oiling Checklist

Harley-Davidson's maintenance employees routinely check other critical items on the Drill and Tapper. Figure 81 shows the log used to record what maintenance has been done and what remains to be done. This sheet is posted on the **Equipment Action Board** near the machine. Notice how it refers to PMs. These preventive maintenance plans are lists of activities that keep the machine running right. While performing the scheduled PMs, unscheduled maintenance may be performed and noted.

Machine Maintenance Log

Brass Tag # 3889   Brother CNC Drill and Tapper
Cell 42011   FL RH Crankcase Mach Station 2

Maintenance Department: Date and initial when activity is done.

Planned Maintenance	Completion Dates/Initials
Brother Drill 17-week elec. universal PM	
Brothers Drill 4-week coolant PM	
Brothers Drill and Tapper 4-week filter PM	
Brothers Drill and Tapper 4-week oiling PM	
Brother Drill and Tap 13-week oiling PM	
Brother 4-week clean grip covers PM	
Brother Drill 26-week mech. universal PM	

Unplanned Maintenance

Date	Initials	Activity

Figure 81 – Harley-Davidson Maintenance Log

Daily Operating or Inspection Checklists

Typically, checklists designed in the office can lead to much confusion on the plant floor. To maintain critical readings on a customer's new compressor during the warranty period, a compressor manufacturer required the Daily Log Sheet shown in Figure 82.

BELLISS & MORCOM AIR / GAS COMPRESSOR DAILY LOG SHEET										
HAMWORTHY Bellis & Morcom Aftermarket										
DATE										
COMP TYPE										
SERIAL No.										
TIME READINGS TAKEN										
FIRST STAGE DELIVERY PRESSURE	(BARG)									
SECOND STAGE DELIVERY PRESSURE	(BARG)									
THIRD STAGE DELIVERY PRESSURE	(BARG)									
OIL PRESSURE	(BARG)									
FIRST STAGE INLET TEMPERATURE	(°C)									
FIRST STAGE OUTLET TEMPERATURE	(°C)									
SECOND STAGE INLET TEMPERATURE	(°C)									
SECOND STAGE OUTLET TEMPERATURE	(°C)									
THIRD STAGE INLET TEMPERATURE	(°C)									
THIRD STAGE OUTLET TEMPERATURE	(°C)									
WATER INLET TEMPERATURE	(°C)									
WATER OUTLET TEMPERATURE	(°C)									
WATER FLOW RATE	(GPH)									
CURRENT L1 AMPS										
CURRENT L2 AMPS										
CURRENT L3 AMPS										
VOLTS										
P.F.										
KW										
HOURS RUN:	0 LOAD									
	½ LOAD									
	FULL LOAD									

<D&G Belliss Log Sheet.doc>

Figure 82 – Compressor daily log sheet

Many errors were being reported by the engineers when operators and technicians recorded the data. The form was **not as user-friendly** on the plant floor as it was in the engineer's office. By applying some basic visual principles to the checklist, it was transformed as shown in Figure 83.

Desnoes & Geddes Ltd.
Belliss & Morcom Compressor Shift Log and Inspection Sheet

Date: _____   Shift: _____

No	Checks (see "Operator PM Instructions")	Units	Normal Reading	Actual Reading On Load	Actual Reading Off Load
	PRESSURE				
1	Oil Pressure	BAR	2 to 6		
2	1st Stage Air Pressure	BAR	3 to 3.5		
3	2nd Stage Air Pressure	BAR	14 to 15.5		
4	Final Air Delivery Pressure	BAR	40 to 45.5		
5	Air Receiver	BAR	40		
	COOLING				
6	1st Stage Cyl. & Intercooler	M³/h	5.5 to 6.5		
7	2nd Stage Cyl. & Intercooler	M³/h	6		
8	After Cooler	M³/h	7		
	AIR TEMPERATURE				
9	1st Stage Delivery	°C	40		
10	2nd Stage Delivery (opposite side) ⑩	°C	30		
11	After Cooler (opposite side) ⑪	°C	25		
	OTHER				
12	Ammeter	Amps	< 410		
13	Crankcase Oil Level (30W)	---------	Half Glass		
14	Test Panel Lamp Operation	---------	All Light		
15	Check Load/Unload Operation	---------	Unloads		
16	Hour Meter at Start of Shift	Hours			

Locations — Left Side of Compressor

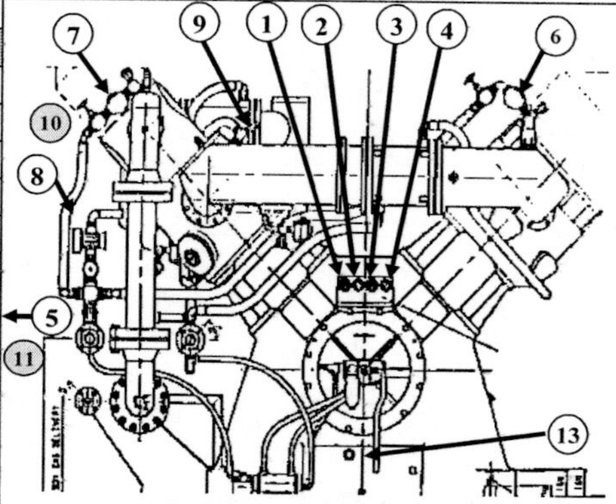

Ammeter & Hour Meter on Control Panel

Remarks:

(continue on reverse side if needed)

Signatures Dates & Routing:
- Operator:
- Supervisor/Team Leader:
- Maintenance Engineer:
- Equipment History File:

Figure 83 – Revised "Visual" Compressor Log Sheet

Visual Checklist

Figure 82 shows how all the inspection points are included and grouped. Extra inspection points were deleted from the checklist. A pictorial was added along with two digital photos to show where each reading was to be taken. A routing/signoff block at the bottom made sure the right people looked at the information in order to take corrective action as needed.

Visual Machine

With elements of error proofing, the next step made this inspection checklist even more user-friendly. The compressor was labeled and numbered exactly as shown on the checklist. This is demonstrated in Figure 84.

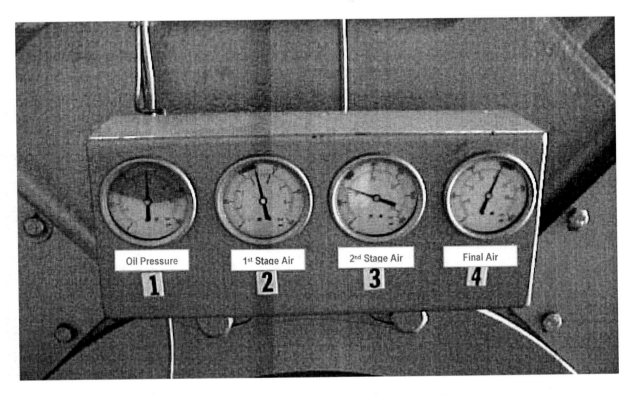

Figure 84 – Compressor inspection labeling

In this application, we have used the following methods:

- Checklist forms with machine-specific visuals
- Equipment inspection point identification: points numbered with adhesive labels, points labeled with names to match list
- Gauges marked with green polyester labeling to show proper operating range
- Gauges marked with red to show high or danger ranges
- Green paint pen marks to show desired operating point on the gauge

Cleaning and Inspection Checklists

A Cleaning and Inspection Checklist is a reminder for what needs to be attended to on a specific piece of equipment. These checklists typically have a drawing or a digital photo of the sides of the machines and the inspection and cleaning points are labeled. Each point is then described along with the type of cleaning and/or inspection step that is required. An example of a Cleaning and Inspection Checklist for an injection molding machine is shown in Figure 85.

Group Name:	Cell 3 Teams A, B, C	Date: 26 Oct	
Machine Name or Area:	CE 33 Back Side	Page:1 of 2	Pages

Draft Cleaning and Inspection Procedures

Insert a photograph or sketch a diagram of the equipment.

Point	Description	Standard	Time	Frequency
1	Oil Filter Pressure Gauge	Clean and check 120 psi	2 minutes	Shift
2, 3, 4, 5, 6	Lube Points	Clean and check for leaks	2 minutes	Weekly
7	Hydraulic Unit	Clean and check for leaks	1 minute	Weekly
8	Chilled Water Pump	Clean and check for leaks / Validate pressure at 45 psi	2 minutes	Shift
9	Chilled Water Pipes	Clean and check for leaks	2 minutes	Weekly
10	Floor	Clean and check for spills & leaks	5 minutes	Shift

Figure 85 – Cleaning and Inspection Checklist

Another example of an inspection checklist is shown in Figure 86. In this case, it is for a small portion of a chemical plant. Notice how each point is labeled in the photograph and separate photographs have been combined onto a single checklist.

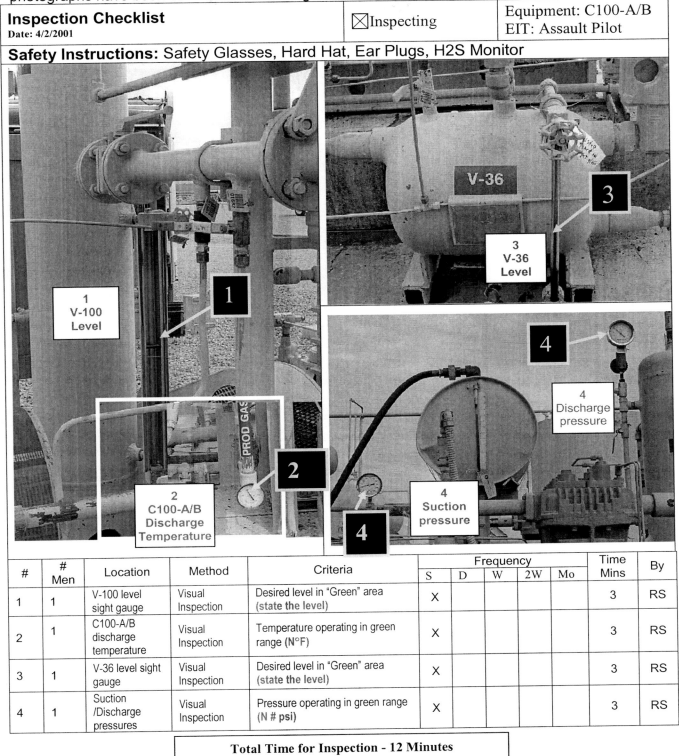

Figure 86 – Inspection Standards Checklist

Our last example is a lubrication procedure for an engraving milling machine. Digital photographs help show where each lube point is located. The table in Figure 87 lists each lube point, who is responsible (operators or mechanics), a brief description of the lube point, its location, the type of lube (color coded), and instructions. This procedure is laminated and posted on the cabinet next to the machine. The PM procedures from the computerized maintenance management system (CMMS) must reinforce the lubrications shown in this procedure, as well as serve as a reminder to perform the lube tasks and document them upon their completion.

Draft Lubrication Procedure - Engraving machines

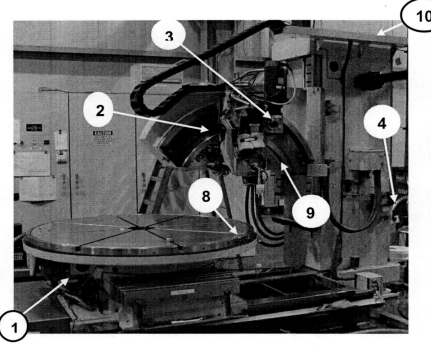
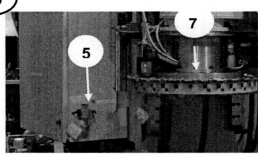

Job	Who	Freq.	Description	Location	Oil	Instructions
1	Oper	W	Table gear box	Under table, left side	H-15	Fill site gage 1/2 full
2	Oper	W	Contrast groove head	A-axis, left side	H-103	Fill to line
3	Oper	W	A-axis gearbox	A-axis, right of spindle	H-15	Fill site gage 1/2 full
4	Oper	W	Way lube	Column, next to control	H-16	Take container to lube room and refill
5	Oper	W	Carousel motor lube	Panel, next to carousel	H-103	Special lube can (TBD)
6	Oper	W	Main hydraulic tank	Hydraulic unit, behind machine	H-11	Fill site gage 3/4 full
7	Oper	Q	Ring gear	Tool carousel, teeth of outer gear	G-01	Brush grease, light coating
8	Maint	Q	Table	T-slot stamped "0" near OD	H-16	(Quantity TBD)
9	Maint	A	A-axis spindle chain	A-axis, underside, behind covers	H-49	Brush lube on chain
10	Maint	A	Z-axis CB chain	Column, top of machine	H-49	Brush lube on chain

Figure 87 – Draft Lubrication Procedures

Chapter 8 Activities and points to ponder…

1. Inspection checklists help make sure that all the right tasks get done:

   ____ True
   ____ False

2. What kinds of checklists are currently used in our plant and on our targeted critical equipment?

   a. _____ Preventive Maintenance (PM)
   b. _____ Lubrication (Lube)
   c. _____ Cleaning
   d. _____ Inspection
   e. _____ Setup and changeover
   f. _____ Sanitizing
   g. _____ (other)

3. What types of checklists could be developed for our targeted critical equipment?

4. What benefits could be realized by improving on our use of checklists?

Products described in this chapter available from www.swspitcrew.com:

\# 036 – 5/8" adhesive all weather numbering labels
\# 037 – 1" adhesive all weather numbering labels
\# 038 – 2" adhesive all weather numbering labels
\# 016 – Gauge Marking Labels
\# 004 – Paint Pens

Pit Crew

© 2006-2012 Strategic Work Systems, Inc.

How should the Oil Spot Test results be interpreted?

Take a look at the oil spot testing examples in Figure 96. Two of the most common dispersant patterns are shown in cards labeled 1 and 2; these indicate an acceptable level of dispersancy and cleanliness. Cards labeled 3 and 4 show unsatisfactory dispersancy. An increase in engine oil deposits can be expected if these conditions exist.

The oil spot test is not designed to replace a thorough oil analysis and/or changing program, but serves as a good indicator of some of the most important characteristics of engine lubricating oils.

Example 1. Very high dispersancy and low insoluble contaminants. The spot is generally light with a dark center and an indistinct periphery.

Example 2. High dispersancy with moderate insoluble contaminants. The pattern shows a relatively light center with a fuzzy band at the periphery.

Example 3. Dispersancy is impaired because of water and/or anti-freeze. The spot shows a sharp periphery (indicating water) and a uniform dark center (indicating anti-freeze).

Example 4. Dispersancy is absent because of attrition. The pattern shows a sharp periphery and a uniform dark field. Density of the pattern depends upon the amount of contamination.

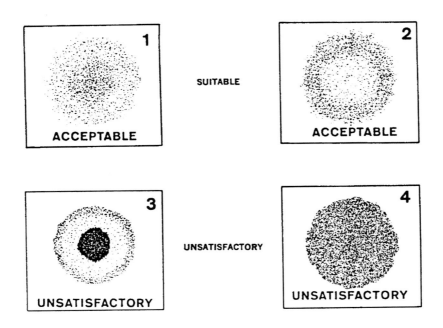

Figure 96 - Oil card testing examples

Chapter 9 Activities and points to ponder…

1. What condition monitoring methods or tools are currently used in your plant?

2. Is there a benefit to more people paying attention to the condition of your critical equipment? Specify.

3. Did you know that other names for "condition monitoring" include…

    a. Condition-based maintenance
    b. Predictive maintenance

4. Which of the condition monitoring methods shown in this chapter would benefit our targeted critical equipment? How?
    a. _____ Temperature sensing labels
    b. _____ Vibration analysis pickup discs and labels
    c. _____ Visual oil test cards
    d. _____ (other)

Products described in this chapter available from www.swspitcrew.com:

\# 006-1 – Temperature sensing strips (170° - 240°F)
\# 006-2 – Temperature sensing strips (130° - 190°F)
\# 032, 033, & 035 – Vibration analysis pickup discs and labels
\# 058 – Preventive Maintenance labels
\# 023 – Visual Oil Test Cards (package of 5 cards)
\# 023 – Paint Pens

CHAPTER 10 - Stored Rotating Equipment

Shaft Targets for Motors, Gearboxes, and Fans

Motors, gearboxes, fans, and other rotating devices often deteriorate during storage and have to be replaced a few weeks after installation. Vibrations in the floor of the storage area are transmitted to the bearings and cause small indentations in the bearing race. These indentations ruin the bearing. This is known as "false Brinelling" (shown in Figure 97), and it can be easily avoided by rotating the shaft on a regular basis.

Figure 97: False Brinelling in new bearing race

Some companies have their larger motors hooked up with bicycle chains and a 1 RPM motor to keep them moving at a constant rate to avoid false Brinelling in bearings and armature sag in motors. There is an easier way. Rotating the shafts by hand is tedious, but it pays big dividends in storage areas that are near traffic areas, running equipment, or other sources of vibration. But how do you know when it is time to rotate the shaft? A small patterned or color-coded disk stuck to the end of the shaft can be used to simplify the preventive maintenance of rotating equipment while in storage. Double-stick tape works well. An example shaft target is shown in Figure 98.

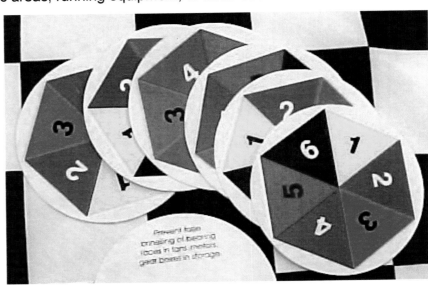

Figure 98: Shaft target

How to Use the Shaft Target

The motor shaft targets should generally be used on any motor of 50 HP or higher. You may also find some other smaller motors that have premature bearing failures after prolonged storage in your facility. This would be reason to apply the shaft targets and rotation to smaller motors. They also should be used on fans and any other large rotating equipment that sits in storage for lengthy periods of time.

By rotating these shafts monthly (or more depending on the floor vibrations), you are preventing armature and shaft sag, as well as false Brinelling of the bearings into the races. By rotating the shaft several times, the oil film is re-distributed on the rolling elements and the inner and outer bearing races.

Set up a shaft rotation schedule to suit your environment. Use data on premature bearing failures on newly installed components. In the absence of such data, evaluate your storage areas. For example, on offshore oil and gas platforms, the entire platform can vibrate with all of the rotating equipment on board. In this case, a rotation of 7- to 10-days may be appropriate.

Some motors, gearboxes, and fans are stored near loading docks or where there is considerable forklift traffic. Monthly rotation of the shafts may be sufficient.

Make a shaft rotation chart and hang it on the wall in the equipment storage areas. Figure 99 shows an example of a *Shaft Rotation Chart*. Indicate the color of the month (or week) and instructions on how to rotate the shafts. The people in charge of stored rotating equipment rotate the shafts two or three times and stop with the required color at the top, or at the 12 o'clock position.

Figure 99 - Shaft rotation schedule

Chapter 10 Activities and points to ponder…

1. Look in your store room. How are the motors, fans, and gearboxes stored?

2. Are any of the motors stored standing on the ends? Are they designed to run in this position? If not, have there been cases of bearing failures after installation of the stored motors?

3. Are the shafts of the motors, fans and gearboxes being rotated periodically as part of a formal maintenance of stored equipment? If yes, who is responsible for assuring the shafts are being rotated on schedule?

4. Search the maintenance work order history. Look for examples of motor, fan, or gearbox failures within several weeks after installation. Could these premature failures be caused by improper storage and lack of maintaining stored spares?

5. Could premature failures be caused by damage received during the transportation of the motor, fan, or gearboxes from the manufacturer, supplier, or vendor to your plant? Are they properly positioned for transportation? Are they inspected upon receipt?

Products described in this chapter available from www.swspitcrew.com:

024 – Motor shaft Targets

CHAPTER 11 - Tools, Parts, and Supplies Inventory Control

Organized and orderly tool boxes, parts, and supply cabinets save significant amounts of time. The more organized they are, the less time is spent hunting for the right item. Here are a few ideas for organizing tools, parts, and supplies.

Hand Tools

NASCAR race teams are known for their fast and accurate work. Part of their secret is the way they organize their tools so they do not have to hunt for anything. Every tool has its place. The following figure shows a tool set being used to do bench work in a NASCAR shop.

Figure 100 – NASCAR mechanic's working tool set

Watching the mechanic work and how he used his tools was very revealing. Rarely did the mechanic place a hand tool anywhere but in the proper spot in the tool tray. That way, he never had to look anywhere but where the tool was supposed to be in the tray.

In one plant, trimmer knife changing was delayed because too much time was spent looking for the right tools and parts. They took a lesson from the NASCAR mechanics and developed a portable tool cabinet with spaces for the right tools. They developed the tool board using a lockable Snap-on® tool cabinet, made a tool shadow board outlining each tool, and mounted it on rollers so it could be easily moved next to the machine where the trimmer knives were to be changed. The only people who had a key to the lock were those who were trained and qualified to change trimmer knives. This simplified "tool box" eliminated clutter, kept the right tools and parts at the right place, and secured them. Figure 101 shows this point-of-use tool storage.

Figure 101 – Point-of-Use Work Area

Grease Guns

Standardizing grease guns in your plant is an important part of proper lubrication. It helps prevent over- and under-greasing and improper grease application, and it improves safety. Different grease guns, made by different manufacturers, supply different amounts of grease with each pump. For proper lubrication, the number of pumps required to deliver an ounce of grease is important. Also, knowing the actual amount of grease delivered per pump may be important. Consult the catalogs and the instruction sheet that comes with the grease gun. The following table shows the differences.

Cartridge/Bulk Fill Lever Operated Grease Gun Volume Table

Grease Gun Model	Operating Pressure	Pumps per Ounce
Lubrimatic 11142	10,000 psi	12
Alemite 500, 500E	10,000 psi	21
Lincoln 1142, 1148, 1147	6,000 psi	33
Legacy L1000-GRA	6,000 psi	39
Lincoln 1037	3,000 psi	9

"Calibrate" your grease guns. Grease guns outputs vary as you can see in the previous table. Standardize to one type of gun from one manufacturer to eliminate this variable that can have a detrimental affect on bearings, pumps, and motors. Apply labels to the grease gun that identify the type of grease, the color code that matches fittings and labeled points in the lube plan, the grams/pump and pump per ounce plus a "calibration due date." Check the grease output volumes at least annually to start.

The Grease Gun Sleeves, as shown in Figure 102, are used to protect the identification, grease type, and color-code labeling applied to the barrel of a grease gun. They are a heat shrinkable polyolefin that is flame retardant with excellent electrical, chemical, and physical properties including resistance to grease, penetrating oil, chlorinated cleaners, sunlight, moisture, and fungus. Temperature range for continuous operation is from -55°C to 135°C. Electrical: 1300 volts per mil shrinks the sleeve to 1.5 inches in diameter and to .050 inches wall thickness.

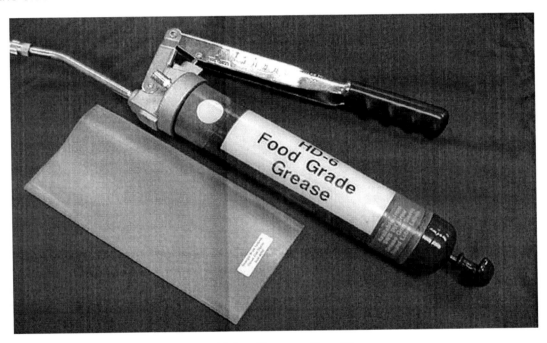

Figure 102 – Grease Gun Sleeve

Directions for Grease Gun Sleeve Use

1. Remove the pump mechanism, grease, and plunger from the grease gun.
2. Clean the outside barrel of the grease gun with a clear solvent and dry thoroughly.
3. Apply desired grease label or color-code labels to barrel of grease gun.
4. Slide the shrink tubing over the barrel and center lengthwise.
5. Using a high-temperature heat gun or infrared heater set at 90°C -194°F, heat the shrink tubing evenly all over. Concentrate heated air along the length until it starts shrinking.
6. Move the heat gun along the shrink tubing until it has shrunk to a smooth surface on the barrel of the grease gun. This may take up to five minutes of heating.
7. Use proper safety measures to prevent being burned during this process (i.e. gloves).

Parts and Supplies

Parts and supplies in a warehouse or on the plant floor can be organized for fast access and to eliminate running out of needed supplies. Figure 103 shows a typical supply storage area that can be organized with pictorial labels. These pictorial labels (**Stock Inventory Cards**) indicate the supplies by name and a visual arrangement on the shelves.

Figure 103 – Supplies storage and labeling

In the stack of supplies on top of the minimum inventory level is a **Stock Reorder Card**. This card is a visual cue that more trolley wheels should be ordered. It specifies the type, the source, minimum level, and lot size to re-order, and the delivery lead time. Figure 104 shows a sample Stock Re-order Card.

Figure 104 – Stock Re-order Card

The conveyor parts stored in the conveyor repair area are easily accessible to the mechanics making the repairs. "Stock outs" are prevented by using a Stock Re-order Card, placed on top of the box with the "minimum level" of parts. This card triggers the re-order of conveyor parts.

Here are some examples of production supplies stock reorder and stock use cards in Figure 105. The **Stock Reorder Card** is placed on top of the last five rolls of tape on the shelf ("Minimum stock"). This reminder card makes reordering easy.

Stock Re-order Card

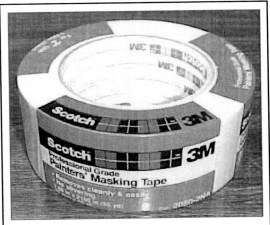

2 inch x 200 ft Masking Tape
Stock # 123345
Cost $4.25 ea - Central Stores
Minimum stock: 5 rolls
Order quantity: 12
5 day delivery

Stock Use Card

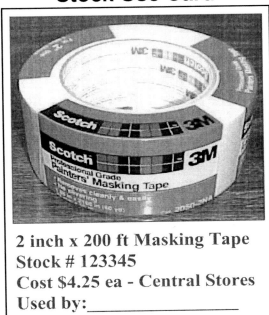

2 inch x 200 ft Masking Tape
Stock # 123345
Cost $4.25 ea - Central Stores
Used by:_____
Date: _____

Figure 105 - Stock reorder and stock use cards

The **Stock Use Card** in Figure 105 keeps track of the quantity and use rate of parts and supplies. As rolls of tape are put on the shelf for use, a card is attached. As the tape is taken from the shelf, the **Stock Use Card** is filled out and dropped in the inventory box.

Oil Sump Sight Glasses

These durable sight glasses shown in Figure 110 are ideal for spotting contaminating sediment and moisture contamination in lube oil sumps and bearing housings in pumps.

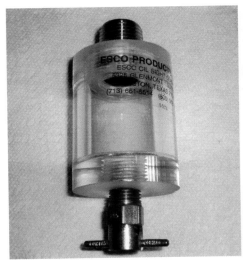

Crystal clear Oil Sight Glass makes it easy to observe bearing oil contamination from dirt, grit and/or water.

Figure 110 – Pump oil sight glass

Esco Oil Sight Glasses are available from:
Esco Products, Inc.
Houston, TX 77081
(800) 966-5514 or (713) 661-5514
www.escopro.com

CREATE YOUR OWN
Lean Machine Visuals
ON-SITE AND ON-DEMAND

Brady software and printers are essential tools for implementing a visual workplace, making it easy for you to create durable, professional-looking signs, labels, tags and more.

Brady printing systems can also create visuals used in lean manufacturing initiatives such as **5S, Standard Work, Quick Changeover and Just-In-Time Inventory Control.**

© 2006-2012 Strategic Work Systems, Inc.

WITH BRADY
Industrial Label Printers

BBP®31 Sign and Label Printer

The BBP®31 Sign and Label Printer is extremely easy to use – no training required! This monochrome label printer offers a tilt-adjustable touchscreen and a fold-down keyboard. Prints from ½" to 4" wide on a variety of specialty materials, including magnetic tape, repositionable labels, tagstock, glow tape and more!

GlobalMark® 2 Industrial Label Maker

The GlobalMark®2 Industrial Label Maker prints multi-color 5S signs and labels from ½" to 4" wide. The GlobalMark®2 printer includes a built-in plotter that cuts out letters, shapes and custom-sized labels. Prints on a wide variety of specialty materials, including floor tape, repositionable labels, tagstock, magnetic tape and more!

MarkWare™ Lean Tools Software

MarkWare Lean Tools works with Brady's label printers, making it even easier for you to create a visually instructive workplace. MarkWare provides versatile drawing tools and hundreds of premade templates for signs, labels and tags used in lean production, maintenance, warehouse and safety applications.

Visual TPM eLearning course
Enhancing Equipment Reliability and Maintenance Efficiency

The eLearning Visual Total Productive Maintenance (TPM) course provides an overview and demonstration of how the use of visual controls can enhance equipment reliability and improve maintenance efficiency. The course includes an overview of the 5 pillars of TPM, best practices for process and equipment identification, use of visuals for maintenance inspections and more!

The course includes:

- Overview of the 5 pillars of TPM
- How to use visuals to facilitate preventative and predictive maintenance inspections
- Visual inventory control methods for storerooms and tool cribs
- Best practices for process and equipment identification
- Tips for creating TPM schedules, procedures and checklists

Online Resource for Visual Workplace Ideas & Solutions

Go online to Brady's visual workplace website for additional application ideas, best practices, articles, webcasts and more! The website also provides further information on all Brady's innovative visual workplace solutions.

www.BradyID.com/visualworkplace

www.BradyID.com/visualworkplace

WHEN PERFORMANCE MATTERS MOST

Insights about Lean Manufacturing and TPM

Throughout this book, I have mentioned "TPM" and "Lean" – two very often misunderstood terms in today's manufacturing plants and facilities. The more time I spend with these two improvement strategies in plants and facilities of all types all over North America, the more I've come to recognize "real TPM" (Total Productive Maintenance) and "real Lean" when I see it.

Too many firms launch their TPM and Lean initiatives as "programs" and "tools." The developers of TPM and Lean, formerly known as the Toyota Production System and Just-in Time Manufacturing, had a very simple underlying principle for both…

"Systematic identification and elimination of waste to reduce manufacturing cost"

This translates to a very simple 10-step improvement process:

- Step 1 – Identify the biggest losses or problem in your business
- Step 2 – Quantify the magnitude of the losses
- Step 3 – Gather data to define the losses
- Step 4 – Establish a hypothesis, a "what if we do this" idea *(ie. Visuals & minor modifications)*
- Step 5 – Conduct an Experiment. Try the idea
- Step 6 – Measure the results of the experiment: Worked? Didn't work?
- Step 7a – Worked: Implement what worked and monitor the progress
- Step 7b – Didn't work: Go back to Step 3 and incorporate the new data
- Step 8 – Set clear expectations and hold everyone accountable for the new methods
- Step 9 – Reinforce the desired behaviors
- Step 10 – Go back to Step 1 and do it all again

While this may be an oversimplification of TPM and Lean, it **is** the underlying principle. True "Lean" is based on the ***Scientific Method*** (Steps 1 – 7b) supported with ***Culture Changing Leadership*** (steps 8 – 10).

"History tells us why things are the way they are today…" **learn from it.**

In the nearly 20 years I have been a student of TPM and the roots of "Lean" I have collected a number of "must-read" **historical** books and articles on the subjects. I offer them to you for your education and application. Avoid the trap of **implementing** the TPM and Lean tools in the **hopes** that you will create a lasting improvement – it rarely happens. Anecdotal data tells us that about 98-percent of 5S (Workplace organization and orderliness) programs have gone dormant or died after 18 months. Why? Chances are they were just another "program."

Suggested Lean and TPM "Historical" Reading List:
- Ford, Henry; *Today and Tomorrow*, 1926 (reprinted). Productivity Press
- Dennis, Pascal; *Lean Production Simplified*, 2002. Productivity Press
- Monden, Yasuhiro; *Toyota Management System – Linking the seven Functional Areas*, 1993 . Productivity Press
- Nakajima, Seiichi; *Introduction to Total Productive Maintenance*, 1984. 1988Productivity Press (out of print) See Note 1
- Nakajima, Seiichi; *TPM Development Program*, 1989. Productivity Press (out of print) See Note 1
- Ohno, Taiichi; *Toyota Production System*, 1978/1988. Productivity Press
- Shingo, Shigeo; *A Study of the Toyota Production System*, 1981. Productivity Press
- *Harvard Business Review*, Sept-Oct 1999; "Decoding the DNA of the Toyota Production System," Kent Bowen and Steven Spear

Additional related reading:
- Maintenance Technology Magazine April 2000. "*TPM: An Often Misunderstood Equipment Improvement Strategy*" Robert M. Williamson
- Goldrattt, Eliyahu M. *The Goal, a Process of Ongoing Improvement* (revised edition), Croton-on-Hudson. New York: North River Press.
- Kotter, John P.C.; *Leading Change*, 1997. Harvard Business School Press
- Schaffer, Robert; *The Breakthrough Strategy*, 1988. Harper Business
- Suess, Dr. (Theodore S. Geisel, et al) (1955). *On Beyond Zebra*. New York: Random House.
- Womak, James and Jones, Daniel; *Lean Thinking*, 1996 & 2003. Simon and Schuster
- Womak, James, Jones, Daniel, and Roos, Daniel; *The Machine That Changed The World*, 1990. Harper Collins

Note 1:
The original Total Productive Maintenance (TPM) books that provided the most comprehensive overview of the design and intent of the equipment improvement methodologies were those written by Seiichi Nakajima. They are a MUST READ for the true TPM Leader in today's **Lean Machine** transformation. However, they are both unfortunately out of print.

We have a number of sets of Nakajima's two TPM books listed above available for purchase. Contact us for price and availability (RobertMW2@cs.com). Supply is limited. If you have some you wish to sell, please contact us also.

RMW

Foreword

Hello and welcome to what's going to be a great learning experience this semester. We are Jerry Shawver and Bill Meisel, the authors of this textbook, and we want to share some of the pedagogical features of the textbook and course with you.

This textbook is aligned with the new online software MathXL. You will notice that the designated number of each of the assignments in MathXL corresponds to the same chapter and section numbers in the textbook. For example, Assignment 1-1 corresponds to Chapter 1, Section 1 in the text, Assignment 1-2 with Chapter 1, Section 2, and so forth. This alignment should make finding explanations easy when you work online.

A major revision in this edition is the addition of more explanations. In the first edition, our intent was to make our explanations as easy as possible for you to understand, and in this edition we feel the explanations are even more understandable. Students who read the previous edition commented on its readability (we try write as if we are speaking to you in our classroom) and stated that this was the first mathematics book that they were able to read and understand. If you read the textbook prior to solving the problems in each chapter, we think you will find the work much easier to complete.

The MathXL software has several features that will help you to complete your work. Features such as "Show Me" and "Guided Solution" benefit most students. Some of the problems even have a video associated with them. Even though the videos don't feature us, we highly recommend them to you.

Enjoy the semester,

Jerry Shawver

Bill Meisel

Chapter 1
Algebra Basics

Assignment Checklist

What You Should Do	Where?			When?	✓
Read the Syllabus and Student Orientation		💻		Prior to Week 1	
Introduce yourself in the Discussion Board		💻		Week 1	
Register in MathXL			MathXL	Week 1	
Read, view the videos, and then complete the online work for Chapter 1, Section 1	📖	💻	MathXL	Week 1	
Read, view the videos, and then complete the online work for Section 2	📖	💻	MathXL	After completing Chapter 1, Section 1	
Read, view the videos, and then complete the online work for Section 3	📖	💻	MathXL	After completing Section 2	
Take the quiz on Chapter 1			MathXL	After completing Section 3	
Post questions and respond to other students' questions in the Discussion Board		💻		Anytime	
Other assignments:					
Notes:					

Section 1
Numbers and Sets

Learning Objectives

When you finish your study of this section, you should be able to
- Classify different types of numbers
- Identify and use various set notation symbols
- Write sets in set notation and set builder notation
- Perform basic operations with sets

Number Sets

N is the set of **natural numbers**. This set includes the numbers we count with every day (no fractions or decimals).

$$N = \{1, 2, 3, 4, 5...\}$$

W is the set of **whole numbers**. This set includes the natural numbers plus 0.

$$W = \{0, 1, 2, 3, 4, 5 ...\}$$

Z is the set of **integers**. This set includes the natural numbers, their opposites, and 0.

$$Z = \{...-4, -3, -2, -1, 0, 1, 2, 3, 4...\}$$

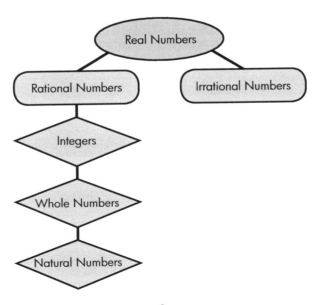

$-3 - 11$ Rewrite as an addition problem.

$-3 + (-11)$ Add the numbers. Keep the sign.

Answer: -14

Example 12

Simplify: $-5 - (-2)$

$-5 - (-2)$ Rewrite as an addition problem.

$-5 + 2$ $5 - 2 = 3$. Since 5 is negative, the answer is negative.

Answer: -3

Example 13

Simplify: $10 - (-4) + (-7)$

$10 - (-4) + (-7)$ Rewrite as an addition problem.

$10 + 4 + (-7)$ Add the two positive numbers first.

$14 + (-7)$ $14 - 7 = 7$. Since 14 is positive, the answer is positive.

Answer: 7

Multiplication and Division of Real Numbers

When multiplying or dividing real numbers, pay attention to the number of negative signs you see in the problem. If the number of negative signs is **even**, the answer is positive. If the number of negative signs is **odd**, the answer is negative. That's all there is to it.

Example 14

Simplify: $(-3)(-2)(4)$

$(-3)(-2)(4)$ Two negative signs: the answer is positive.

Answer: 24

Example 15

Simplify: $(-1)(-5)(-7)(2)$

$(-1)(-5)(-7)(2)$ Three negative signs: the answer is negative.

Answer: -70

Chapter 1: Algebra Basics

Example 16

Simplify: $\left(\dfrac{-6}{5}\right)\left(\dfrac{2}{3}\right)$

$\left(\dfrac{-6}{5}\right)\left(\dfrac{2}{3}\right)$

$\dfrac{\overset{-2}{\cancel{-6}}}{5} \cdot \dfrac{2}{\cancel{3}}$ $\qquad$ $-6 \div 3 = -2$.

$\dfrac{-2}{5} \cdot \dfrac{2}{1}$ $\qquad$ Multiply the numerators. Multiply the denominators.

Answer: $\dfrac{-4}{5}$ $\qquad$ One negative sign: the answer is negative.

Example 17

Simplify: $(-2.34)(-.02)$

When multiplying decimals, just multiply the numbers in the usual way. There are two negative signs, so the answer is positive. $234 \times 2 = 468$. There are **two** decimal places to the right of -2.34 and **two** decimal places to the right of $-.02$, so there should be **four** decimal places to the right of the final answer.

Answer: 0.0468

Note: You might use a calculator on this one.

Example 18

Simplify: $(-8) \div (-2)$

$(-8) \div (-2)$ $\qquad$ Two negative signs: the answer is positive.

Answer: 4

Example 19

Simplify: $\left(-\dfrac{4}{5}\right) \div \left(-\dfrac{3}{5}\right)$

$\left(-\dfrac{4}{5}\right) \div \left(-\dfrac{3}{5}\right)$ $\qquad$ To divide two fractions, change to multiplication and flip the second fraction over.

$\dfrac{-4}{5} \cdot \dfrac{-5}{3}$

$\dfrac{(-4) \cdot (-5)}{5 \cdot 3}$ $\qquad$ Two negative signs: the answer is positive.

$\dfrac{20}{15}$ $\qquad$ Divide the top and bottom by the common factor, 5.

Answer: $\dfrac{4}{3}$

Example 20

Simplify: $(3.689) \div (-.014)$

$(3.689) \div (-.014)$ Use a calculator.

Answer: -263.5 One negative sign: the answer is negative.

Example 21

Simplify: $18 \div 0$

Let's review division problems involving zero. When 0 is divided by any number, the answer is 0, but a number divided by 0 is said to be "undefined." It doesn't make sense to divide a number by 0, so we call such problems undefined. **Remember:** If 0 is the **first** number in the problem, the answer is 0. If 0 is the **second** number in the problem, the answer is undefined.

$18 \div 0$ The second number is 0. The answer is undefined.

Answer: undefined

Practice Problems

Simplify the following:

1. $|-5|$
2. $|6|$
3. $-|-9|$
4. $-|3|$
5. $-9 + 8$
6. $(-3) + (-6)$
7. $(10) + (-12)$
8. $4 + (-4)$
9. $|-7| + (-9)$
10. $|-6| + |6|$
11. $5.35 + (-1.46)$
12. $(-43.24) + (-1.278)$
13. $\left(\frac{6}{7}\right) + \left(\frac{-1}{4}\right)$
14. $\left(\frac{-2}{5}\right) + \left(\frac{-3}{4}\right)$
15. $8 - 15$

16. $-3-5$
17. $(-5)-(-9)$
18. $12-(-20)$
19. $-3+5-4$
20. $-7-6-(-2)$
21. $14-10+(-7)$
22. $15-(-3)-4$
23. $|-5|+|-3|-4$
24. $6-|-7|-|8|$
25. $\left(\frac{1}{2}\right)-\left(\frac{3}{5}\right)$
26. $\left(\frac{-6}{5}\right)-\left(\frac{-1}{3}\right)$
27. $(3.02)-(4.35)$
28. $(1.05)-(-6.1)$
29. $(-3)(5)(-2)$
30. $(-1)(-2)(-3)(-4)$
31. $\left(\frac{9}{7}\right)\left(\frac{-2}{3}\right)$
32. $\left(\frac{-3}{5}\right)\left(\frac{-10}{7}\right)$
33. $(-2.3)(-5.2)$
34. $(1.26)(-2.03)(-1)$
35. $(10) \div (-2)$
36. $(-45) \div (-15)$
37. $(2.12) \div (-.02)$
38. $(-16.435) \div (-.15)$
39. $\left(\frac{8}{5}\right) \div \left(-\frac{1}{5}\right)$
40. $(-4) \div \left(\frac{-1}{4}\right)$
41. $-16 \div 0$
42. $0 \div (-2)$

SECTION 3
SIMPLIFYING ALGEBRAIC EXPRESSIONS

Learning Objectives

When you finish your study of this section, you should be able to
- Simplify algebraic expressions involving exponents
- Simplify algebraic expressions involving square roots
- Use the order of operations rules to simplify algebraic expressions
- Evaluate algebraic expressions through substitution
- Use the distributive property to simplify algebraic expressions
- Simplify algebraic expressions by combining like terms

Exponents

Definition: In the expression 3^4, 3 is called the **base**, and 4 is called the **exponent**. The exponent tells us how many times to multiply the base by itself. Let's look at some examples:

Example 1
Simplify: 3^4

$3^4 = 3 \cdot 3 \cdot 3 \cdot 3 = 81$
Answer: 81

Example 2
Simplify: $(-2)^3$

$(-2)^3 = (-2)(-2)(-2) = -8$
Answer: -8

Example 3
Simplify: $\left(\frac{-1}{2}\right)^2$

$\left(\frac{-1}{2}\right)^2$ Multiply two $\left(-\frac{1}{2}\right)$s together.

$\dfrac{-1}{2} \cdot \dfrac{-1}{2}$ 　　　　　Multiply the fractions.

Answer: $\dfrac{1}{4}$

Example 4

Simplify: -5^2 　　　　　Evaluate the exponent; then, apply the negative sign.

$-5^2 = -(5 \cdot 5) = -25$.

Answer: -25

Square Roots

The symbol $\sqrt{}$ is called the **square root** sign. It asks you to find the number which, when squared (multiplied by itself), equals the number under the square root sign. Let's look at some examples:

Example 5

Simplify: $\sqrt{49}$

$\sqrt{49} = 7$, since $7 \cdot 7 = 49$.

Answer: 7

Example 6

Simplify: $\sqrt{\dfrac{4}{9}}$ 　　　　　Take the square root of the numerator and denominator.

Answer: $\dfrac{2}{3}$

Example 7

Simplify: $\sqrt{-16}$

Any number, when squared, is **nonnegative** (that is, either positive or 0). Thus, there is no number that we can square to obtain -16.

Answer: No Real Solution

Example 8

Simplify: $-\sqrt{16}$

Answer: -4 　　　　　Evaluate the square root; **then,** apply the negative sign.

Order of Operations

We need to make sure that everyone gets the same answer when doing an arithmetic problem. For example, someone might simplify $3 + 5 \cdot 8$ by doing the addition first and then doing the multiplication. Performing the operations in this order would give an answer of 64. You might do the multiplication first and then the addition. Performing the operations in that order would give an answer of 43. This discrepancy is not good!

The order of operations agreement forces everyone to obtain the exact same answer when performing any group of arithmetic operations. The order of operations rules follow:

Step 1: Perform any operations inside parentheses, brackets, absolute values, or square roots.

Step 2: Evaluate any exponents.

Step 3: Multiply and divide in order from left to right.

Step 4: Add and subtract in order from left to right.

Let's look at some examples:

Example 9

Simplify: $6 - 3 \cdot 4 \div 2^2$

$6 - 3 \cdot 4 \div 2^2$	Evaluate the exponent.
$6 - 3 \cdot 4 \div 4$	Multiply, moving left to right.
$6 - 12 \div 4$	Divide, moving left to right.
$6 - 3$	Subtract.

Answer: 3

Example 10

Simplify: $10 \div 5 \cdot 3^3 - 2$

$10 \div 5 \cdot 3^3 - 2$	Evaluate the exponent.
$10 \div 5 \cdot 27 - 2$	Divide, moving left to right.
$2 \cdot 27 - 2$	Multiply, moving left to right.
$54 - 2$	Subtract.

Answer: 52

Example 11

Simplify: $4 - 3(8 - 4 \div 2)$

$4 - 3(8 - 4 \div 2)$	Inside parentheses, divide and
$4 - 3(8 - 2)$	Then, subtract.

4 − 3(6)   Multiply before subtracting.

4 − 18   Subtract.

Answer: −14

Example 12

Simplify: $2[18 \div 3 - (6 - 4)]$

$2[18 \div 3 - (6 - 4)]$ Begin with the innermost parentheses.

$2[18 \div 3 - 2]$ Inside the brackets, divide first.

$2[6 - 2]$ Then, subtract.

$2(4)$ Multiply.

Answer: 8

Example 13

Simplify: $\dfrac{5^2 - 6 \div 2}{3 - 4^3 \div 8}$

In a problem like this, simplify the numerator and denominator separately; then, reduce the fraction to lowest terms if possible.

$\dfrac{5^2 - 6 \div 2}{3 - 4^3 \div 8}$ Evaluate exponents.

$\dfrac{25 - 6 \div 2}{3 - 64 \div 8}$ Divide.

$\dfrac{25 - 3}{3 - 8}$ Subtract.

$\dfrac{22}{-5}$

Answer: $-\dfrac{22}{5}$

Note: Strictly speaking, $\dfrac{22}{-5}$ is not wrong. However, most instructors will prefer placing the negative sign in front of the fraction or in the numerator.

Evaluating Expressions

In these problems, we want to replace the variables with numbers. The best advice we can give you is to change the variables to parentheses first; then, place the appropriate values in the parentheses, and follow the order of operations rules. You will be less likely to make mistakes if you follow this advice. Let's look at some examples.

For Examples 14 through 19, let $a = 2$, $b = -3$, and $c = 4$.

Example 14

Evaluate: $ab + bc$

$(\ )(\ ) + (\ )(\ )$ Change the variables to parentheses.

$(2)(-3) + (-3)(4)$ Place the numbers inside. Multiply.

$(-6) + (-12)$ Add.

Answer: -18

Example 15

Evaluate: $b^2 - 4ac$

$(\ )^2 - 4(\ )(\ )$ Change the variables to parentheses.

$(-3)^2 - 4(2)(4)$ Place the numbers inside. Evaluate the exponent. Multiply.

$9 - 32$ Subtract.

Answer: -23

Example 16

Evaluate: $a^2 - 2a + 5$

$a^2 - 2a + 5$

$(2)^2 - 2(2) + 5$ Simplify the exponent and then multiply.

$4 - 4 + 5$ Add and subtract in order, left to right.

Answer: 5

Example 17

Evaluate: $(b + c)(b - c)$

$(b + c)(b - c)$

$(-3 + 4)(-3 - 4)$ Perform the operation inside parentheses first.

$(1)(-7)$ Multiply.

Answer: -7

Example 18

Evaluate: $|abc| - 5$

$|abc| - 5$	
$|(2)(-3)(4)| - 5$	Multiply inside the absolute value signs.
$|-24| - 5$	Take the absolute value.
$24 - 5$	Subtract.

Answer: 19

Example 19

Evaluate: $\sqrt{ac^2 + b}$

$\sqrt{ac^2 + b}$	
$\sqrt{(2)(4)^2 + (-3)}$	Evaluate the exponent first.
$\sqrt{(2)(16) + (-3)}$	Multiply.
$\sqrt{32 + (-3)}$	Add.

Answer: $\sqrt{29}$

Distributive Property

The **distributive property** states that $a(b + c) = ab + ac$. Our advice is to make sure you multiply the expression on the outside of the parentheses by **each** term inside the parentheses. Let's look at some examples:

Example 20

Simplify: $-4(x - 6)$

$-4(x - 6)$ $(-4)(x) = -4x. (-4)(-6) = +24.$

Answer: $-4x + 24$

Example 21

Simplify: $3(2x^2 - 5x + 1)$

$3(2x^2 - 5x + 1)$ $(3)(2x^2) = 6x^2. (3)(-5x) = -15x. 3(+1) = +3.$

Answer: $6x^2 - 15x + 3$

Example 22

Simplify: $3x(4x^3 - 2x)$

$3x(4x^3 - 2x)$ $\qquad$ $(3x)(4x^3) = 12x^4. (3x)(-2x) = -6x^2.$

Answer: $12x^4 - 6x^2$

Simplifying Expressions by Combining Like Terms

Finally, we look at combining like terms in variable expressions. We follow the basic addition rules when combining the coefficients of the variable expressions.

Recall that two expressions are **like terms** if they have the **same** variables with the **same** exponents. Remember to put your answer in decreasing power order (that is, from the largest exponent down to the smallest exponent).

Example 23

Simplify: $4x^3 - x^2 + 2x + x^3 - 3x^2$

$4x^3 - x^2 + 2x + x^3 - 3x^2$ $\qquad$ $(4+1)x^3 = 5x^3. (-1-3)x^2 = -4x^2.$

Answer: $5x^3 - 4x^2 + 2x$

Example 24

Simplify: $4y^2 + 6xy + y - 4xy + x$

$4y^2 + 6xy + y - 4xy + x$ $\qquad$ $(6-4)xy = 2xy.$ No other terms are like.

Answer: $4y^2 + 2xy + y + x$

Example 25

Simplify: $3(2x - 4) + 6x - 5$

$3(2x - 4) + 6x - 5$ $\qquad$ (**Hint:** Use the distributive property.)

$(3)(2x) = 6x. (3)(-4) = -12$

$6x - 12 + 6x - 5$ $\qquad$ $(6+6)x = 12x. (-12-5) = -17.$

Answer: $12x - 17$

Example 26

Simplify: $5x(x - 4) - (6x - 4)$

$5x(x - 4) - (6x - 4)$ $\qquad$ $(5x)(x) = 5x^2. (5x)(-4) = -20x. (-1)(6x) = -6x.$

$5x^2 - 20x - 6x + 4$ $\qquad$ $(-1)(-4) = +4.$ Then, $(-20 - 6)x = -26x.$

Answer: $5x^2 - 26x + 4$

Practice Problems

Simplify the following.

1. $(2)^4$
2. $(-7)^2$
3. $(-4)^3$
4. $(3)^4$
5. $\left(\dfrac{-2}{3}\right)^3$
6. $\left(\dfrac{-5}{2}\right)^2$
7. -7^2
8. -1^5
9. $\sqrt{64}$
10. $-\sqrt{4}$
11. $\sqrt{-9}$
12. $\sqrt{100}$
13. $\sqrt{\dfrac{36}{25}}$
14. $-\sqrt{\dfrac{49}{81}}$

Use the order of operations to simplify the following.

15. $9 - 3 \cdot 2$
16. $14 + 8 \div 4$
17. $4 \cdot 3^2 - 5$
18. $7 + 2^4 \div 4$
19. $9 - 2 \cdot 6 \div 3$
20. $16 + 5 \cdot 8 \div 2^2$
21. $6 - 4(12 - 3 \cdot 2)$
22. $4 + 3(5 - 2^2 + 1)$
23. $3[15 \div 5(2 - 4)]$
24. $-2[4 \cdot 5(3 - 2^3 \cdot 4)]$
25. $\dfrac{6 - 4^2 \div 2}{3 - 2 \cdot 4}$
26. $\dfrac{16 + 5 \cdot (-3)^2}{6^2 - 4^3}$

Evaluate the following expressions. Let $x = -1$, $y = 2$, and $z = 4$.

27. $x - yz$
28. $y^2 + 3y - 5$
29. $3x + 5z - 7$
30. $z^2 + y^3 - x$
31. $(x - y)(x + y)$
32. $-6(x - 4y)$
33. $|x - 2y| + z$
34. $|2xy - z| - z^2$
35. $\sqrt{x^2 + yz}$
36. $\sqrt{z^2 - 5y + 5x}$

Simplify the following.

37. $-5(3x - 4)$
38. $8(6x^2 - 2x + 1)$
39. $2x(3x - 7)$
40. $-3x^2(5x^2 - 5y + 3)$
41. $x^2 + 3x - 5 + 4x^2 - x$
42. $3x^2 - xy + y^2 + 5xy - 7y^2$
43. $9(x - 2) + 4x - 5$
44. $-2(x - 7) + 3(5x - 8)$
45. $-2(x^2 - 5x + 8) + 7(x^2 - 9)$
46. $x(3x + 2y - 1) - 2y(4x - 3y + 4)$

Chapter 2
Solving Various Equations

Assignment Checklist

What You Should Do	Where?			When?	☑
Read, view the videos, and then complete the online work for Chapter 2, Section 1	📖	💻	MathXL	After completing Chapter 1	
Read, view the videos, and then complete the online work for Section 2	📖	💻	MathXL	After completing Chapter 2, Section 1	
Read, view the videos, and then complete the online work for Section 3	📖	💻	MathXL	After completing Section 2	
Read, view the videos, and then complete the online work for Section 4	📖	💻	MathXL	After completing Section 3	
Read, view the videos, and then complete the online work for Section 5	📖	💻	MathXL	After completing Section 4	
Take the quiz on Chapter 2			MathXL	After completing Section 5	
Post questions and respond to other students' questions in the Discussion Board		💻		Anytime	
Other assignments:					
Notes:					

Section 1
Solving Linear Equations in One Variable

Learning Objectives

When you finish your study of this section, you should be able to
- Solve all types of linear equations
- Identify identities and contradictions

One-Step Equations

First, we will look at equations that can be solved in one step. Recall that the goal is to isolate the variable on one side (usually the left) and a single real number on the other side (usually the right). To do this, we use inverse operations. Addition and subtraction are inverse operations, and multiplication and division are inverse operations. Let's look at some examples.

Example 1

Solve: $x + 6 = 5$

$x + 6 = 5$ Because 6 is **added** to x, you **subtract** 6 from both sides.

$x + 6 - 6 = 5 - 6$ Subtract.

Answer: $x = -1$

Check: $-1 + 6 = 5$. It works!

Example 2

Solve: $x - 3 = 10$

$x - 3 = 10$ Because 3 is **subtracted** from x, you **add** 3 to both sides.

$x - 3 + 3 = 10 + 3$ Add.

Answer: $x = 13$

Check: $13 - 3 = 10$. Right again!

Example 3

Solve: $5x = 20$

$5x = 20$ Because 5 is **multiplied** by x, you **divide** by 5.

Section 2
Literal Equations and Applications

Learning Objectives

When you finish your study of this section, you should be able to
- Evaluate algebraic formulas, given numerical values
- Solve literal equations for the indicated variables
- Solve a word problem, given an appropriate formula

Evaluating Algebraic Formulas

In evaluating algebraic formulas, your goal is to substitute the given values and then simplify using the order of operations rules. Let's look at some examples.

Example 1

Evaluate: $A = \pi r^2$ when $r = 4\ ft$

$A = \pi r^2$ when $r = 4\ ft$ Replace r with parentheses.

$A = \pi(4\ ft)^2$ Let $r = 4\ ft$.

$A = 16\pi\ ft^2$ Multiply.

Example 2

Evaluate: $x = \dfrac{-b \pm \sqrt{b^2 - 4ac}}{2a}$ when $a = 2$, $b = 3$, and $c = -4$

$x = \dfrac{-b \pm \sqrt{b^2 - 4ac}}{2a}$ Replace all variables with parentheses.

$x = \dfrac{-(3) \pm \sqrt{(3)^2 - 4(2)(-4)}}{2(2)}$ Let $a = 2$, $b = 3$, and $c = -4$. Evaluate the exponent. Multiply.

$x = \dfrac{-3 \pm \sqrt{9 + 32}}{4}$ Add under the square root sign.

Answer: $x = \dfrac{-3 \pm \sqrt{41}}{4}$ Since the radical can't be simplified, this is your answer.

Solving Equations for Indicated Variables

In these problems, we follow exactly the same rules that we discussed in the previous section. These problems often seem more difficult to students, though, because variables are in the problems instead of numbers. A good piece of advice is to always try to isolate, on one side of the equation, the variable you are asked to solve for.

Let's try to work through some examples.

Example 3

Solve: $4x + 3y = 10$ for x

$4x + 3y = 10$	Because $3y$ is added, you subtract $3y$ from both sides.
$4x + 3y - 3y = 10 - 3y$	Combine like terms.
$4x = 10 - 3y$	To solve for x, divide both sides by 4.
Answer: $x = \dfrac{10 - 3y}{4}$	You could also simplify this answer.

Simplified answer: $x = \dfrac{10}{4} - \dfrac{3}{4}y = \dfrac{5}{2} - \dfrac{3}{4}y$

Example 4

Solve: $5x - y = 7$ for y

$5x - y = 7$	To eliminate $5x$, subtract it from both sides.
$5x - 5x - y = 7 - 5x$	Combine like terms.
$-y = 7 - 5x$	$-y$ means $-1y$. Divide both sides by -1.
$\dfrac{-y}{-1} = \dfrac{7 - 5x}{-1}$	

Answer: $y = -7 + 5x$ or $5x - 7$

Example 5

Solve: $P = 2l + 2w$ for w

$P = 2l + 2w$	To eliminate $2l$, subtract it from both sides.
$P - 2l = 2l - 2l + 2w$	Combine like terms.
$P - 2l = 2w$	Divide both sides by 2.
$\dfrac{P - 2l}{2} = \dfrac{2w}{2}$	
$\dfrac{P - 2l}{2} = w$	You could also simplify this answer.

Answer: $w = \dfrac{P}{2} - \dfrac{2l}{2} = \dfrac{P}{2} - l$

Example 6

Solve $F = \dfrac{mv^2}{r}$ for r.

$F = \dfrac{mv^2}{r}$ 	Clear the denominator by multiplying both sides by r.

$r(F) = r\left(\dfrac{mv^2}{r}\right)$ 	Simplify.

$Fr = mv^2$ 	To solve for r, divide both sides by F.

$\dfrac{Fr}{F} = \dfrac{mv^2}{F}$ 	Simplify.

Answer: $r = \dfrac{mv^2}{F}$

Solving Word Problems Using Formulas

We will now introduce a few formulas that you can use to solve certain types of word problems.

SIMPLE INTEREST FORMULA

The first formula is the formula for simple interest, $I = Prt$. In other words, this formula says that interest = (principal)(rate)(time), where the principal is the amount of money you start with, the rate is the interest rate paid by the bank, and time is measured in years.

It is worth mentioning that real banks use the compound interest formula to calculate the amount of interest they pay; however, it is traditional to introduce this topic with the less complicated simple interest formula.

Example 7

Suppose you receive a student loan for $10,000 at 5% interest that will be due four years from the time you accept it. If the loan is a simple interest loan, determine the amount of interest you will pay on the $10,000 in four years.

$I = Prt$ 	$P = \$10,000,\ r = 5\% = .05,\ t = 4$ years

$I = (10,000)(.05)(4)$ 	Multiply.

Answer: $I = \$2,000$

In other words, you will pay back a total of $10,000 + $2,000 = $12,000.

COMPOUND INTEREST FORMULA

Now, let's look at the previously mentioned compound interest formula. The formula is $A = P\left(1 + \dfrac{r}{n}\right)^{nt}$, where P is again the principal, r is the rate, t is the time in years, and n is how often the interest is compounded. If the interest is compounded annually, $n = 1$; semiannually, $n = 2$; quarterly, $n = 4$; monthly, $n = 12$; and daily, $n = 365$.

Example 8

Suppose you receive $10,000 from all your family and friends after graduation from college. Instead of spending it, you decided to deposit the funds in a savings account that earns 4% a year, compounded monthly. Use the formula $A = P\left(1 + \frac{r}{n}\right)^{nt}$ to compute the account balance after 10 years.

$A = P\left(1 + \frac{r}{n}\right)^{nt}$ $P = \$10,000, r = .04, t = 10, n = 12$

$A = (\$10,000)\left(1 + \frac{.04}{12}\right)^{(12)(10)}$ Multiply in the exponent. Divide inside parentheses.

$A = (\$10,000)(1 + .00333\overline{3})^{120}$ Evaluate the exponent. Keep all decimal places until the final answer.

$A = (\$10,000)(1.4908326)$

Answer: $A = \$14,908.33$ Round to dollars and cents.

Practice Problems

Evaluate each formula using the given variables.

1. $I = Prt$ Find I when $P = \$500$, $r = 6\%$, and time is 3 years.
2. $I = Prt$ Find P when $I = \$60.00$, $r = 3.5\%$, and time is 4 years.
3. $F = \frac{mv^2}{r}$ Find F when $m = 10$, $v = 3$, and $r = 2$.
4. $P = 2l + 2w$ Find w when $P = 50\,ft$ and $l = 10\,ft$.
5. $Y = mx + b$ Find b when $Y = 8$, $m = \frac{1}{2}$, and $x = 6$.
6. $X = \frac{-b \pm \sqrt{b^2 - 4ac}}{2a}$ Find x when $b = 3$, $a = 2$, and $c = -4$
7. $X = \frac{-b \pm \sqrt{b^2 - 4ac}}{2a}$ Find x when $b = -1$, $a = 4$, and $c = -2$.
8. $A = \frac{1}{2}h(b_1 + b_2)$ Find A when $h = 3\,ft$, $b_1 = 5\,ft$, and $b_2 = 6\,ft$.
9. $A = \frac{1}{2}h(b_1 + b_2)$ Find b_1 when $A = 64\,ft$, $h = 4\,ft$, and $b_2 = 6.6\,ft$.
10. $C = \frac{5}{9}(F - 32)$ Find C when $F = 90°$.

Solve the following equations or formulas for the indicated variable.

11. $6x + 2y = 8$ for x

12. $-3x + 4y = 9$ for x

13. $5x - 2y = 6$ for y

14. $7x - y = 14$ for y

15. $2(x - 4) - 4y = 10$ for y

16. $4(x + 2) + y = 6$ for x

17. $F = ma$ for m

18. $D = rt$ for t

19. $I = Prt$ for r

20. $C = 2\pi r$ for r

21. $Y = mx + b$ for x

22. $P = 2l + 2w$ for w

23. $V = \pi r^2 h$ for h

24. $A = \dfrac{1}{2} bh$ for b

25. $V = \dfrac{1}{3} lwh$ for w

26. $F = \dfrac{9}{5} C + 32$ for C

27. $A = \dfrac{1}{2} h(b_1 + b_2)$ for h

28. $C = \dfrac{5}{9}(F - 32)$ for F

29. $m = \dfrac{y_2 - y_1}{x_2 - x_1}$ for x_1

Solve the following word problems using the formulas noted in Examples 7 and 8.

30. Suppose you borrowed $4,000.00 at 5% interest for 6 years on a simple interest student loan. How much interest would you accrue at the end of 6 years assuming you had made no payments on the loan?

31. Suppose you accrued $1,080.00 in interest over 3 years on a simple interest loan at 4.5%. Determine the amount of principal borrowed.

32. If you accrued $1,125 in interest in 3 years on a simple interest loan of $2,500, what interest rate did you have?

33. Suppose you invest $5,000 in a 3-year CD that compounds interest quarterly at 4%. How much will you have in the account at the end of the 3 years?

34. How much of an initial deposit is needed to have $50,000$ in an IRA after 10 years with interest compounded monthly at 6%? Assume no other contributions are made.

35. How much of an initial deposit is needed to have 1 million dollars in an IRA after 35 years with interest compounded quarterly at 11%? Again we are assuming no contributions are made during the 35 years.

Section 3
Applications of Linear Equations

Learning Objectives

When you finish your study of this section, you should be able to
- Translate words into algebraic expressions
- Solve consecutive number word problems
- Solve word problems involving perimeter and area
- Solve simple interest word problems
- Solve word problems involving commission

Translating Words Into Algebraic Expressions

Below is a table that lists the words that typically appear in word problems and their mathematical equivalents:

	Key Terms	Example	Translation
Addition	Sum More than Increased by	The sum of a number and two Three more than a number Five increased by a number	$n + 2$ or $2 + n$ $n + 3$ or $3 + n$ $5 + n$ or $n + 5$
Subtraction	Difference Less than Decreased by	The difference of a number and one. Three less than a number. A number decreased by four.	$n - 1$ $n - 3$ $n - 4$
Multiplication	Product Times Twice Squared	The product of a number and six Ten times a number Twice a number A number squared	$6x$ $10x$ $2x$ x^2
Division	Quotient	The quotient of five and a number	$\dfrac{5}{n}$
Equal Sign	Is Is the same as	The sum of a number and one is two. The product of seven and a number is the same as twelve.	$n + 1 = 2$ $7n = 12$

Example 1

Translate: Six less than twice a number is eight.

When you see the words *less than*, seeing them not only means subtraction; it also tells you that the number that comes before the words *less than* comes second in the subtraction problems. This explanation is a complicated way of saying that six less than twice a number means $2x - 6$, not $6 - 2x$. The word *is* corresponds to =, the equal sign.

Answer: $2x - 6 = 8$

Example 2

Translate: The quotient of a number and three is five.

Quotient means division, and *is* corresponds to =.

Answer: $\frac{x}{3} = 5$

Example 3

Translate: Four times the sum of a number and ten is four.

When you see the words *sum* or *difference* with a number being multiplied by it, the **sum or difference must be placed in parentheses**.

Answer: $4(x + 10) = 4$

Consecutive Number Word Problems

The numbers 3, 4, 5, 6, 7 are five **consecutive numbers** because they can be written as 3, 3 + 1, 3 + 2, 3 + 3, 3 + 4. Thus, in a word problem asking you to find three consecutive numbers, we can write the three numbers as x, $x + 1$, and $x + 2$.

Consecutive even numbers 4, 6, 8, 10 and **consecutive odd numbers** 5, 7, 9, 11 can be written as 4, 4 + 2, 4 + 4, 4 + 6 and 5, 5 + 2, 5 + 4, 5 + 6, respectively. Thus, in a word problem asking you to find three consecutive even or three consecutive odd numbers, we can write the numbers as x, $x + 2$, and $x + 4$. Be careful: Three consecutive odd numbers are **not** x, $x + 1$, and $x + 3$.

Example 4

Solve: The sum of two consecutive integers is twenty-one.

Two consecutive integers can be represented as x and $x + 1$.

Add these two numbers and set them equal to 21:

$(x) + (x + 1) = 21$

$2x + 1 = 21$ Drop parentheses and combine like terms.

$2x + 1 - 1 = 21 - 1$ Subtract 1 from both sides.

$2x = 20$

$\dfrac{2x}{2} = \dfrac{20}{2}$ Divide both sides by 2.

$x = 10$

Answer: The two numbers are 10 and $10 + 1 = 11$.

Check: $10 + 11 = 21$

Example 5

Solve: The sum of three consecutive odd integers is thirty-three.

Three consecutive odd integers can be represented as x, $x + 2$, and $x + 4$.

Add the three numbers and set them equal to 33:

$(x) + (x + 2) + (x + 4) = 33$

$3x + 6 = 33$ Combine like terms.

$3x + 6 - 6 = 33 - 6$ Subtract 6 from both sides.

$3x = 27$

$\dfrac{3x}{3} = \dfrac{27}{3}$ Divide both sides by 3.

$x = 9$

Answer: The three numbers are 9, 11, and 13.

Check: $9 + 11 + 13 = 33$

Word Problems Involving Perimeter and Area

For now, we will be interested only in the perimeter and area of triangles and rectangles. The perimeter of a figure is the distance you would travel if the geometric figure was a track that you were running around. The area of a geometric figure is the amount of space on the inside of the figure (imagine you were painting the interior of a triangle or rectangle). Let's look at the formulas for some basic shapes in geometry.

TRIANGLES

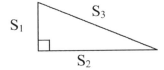

The **perimeter** of a triangle is simply length of side one + length of side two + length of side three. That is, $P = S_1 + S_2 + S_3$. This formula is used for any type of triangle.

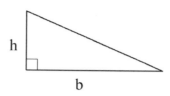 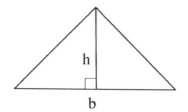

The **area** of a triangle is $\frac{1}{2}$ · base · height. (See pictures above.)
That is, $A = 0.5bh$.

Note: The height is always perpendicular to the horizontal base.

RECTANGLES

A rectangle is a four-sided figure containing four right angles:

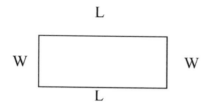

The **perimeter** of a rectangle is simply length + width + length + width = 2 · length + 2 · width.
That is, $P = 2L + 2W$.

The **area** of a rectangle is length · width.
That is, $A = LW$.

OTHER GEOMETRIC SHAPES AND THEIR FORMULAS

Square: Area = s^2.

Parallelogram: Area = bh.

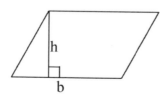

Chapter 2: Solving Various Equations

Trapezoid: Area = $\frac{1}{2}h(a+b)$.

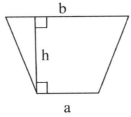

Circle: Circumference = πd or $C = 2\pi r$. (*Diameter = 2r.*)
Area = πr^2.

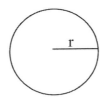

Example 6

John's friend gave him 140 linear feet of fence. He has been contemplating erecting a fence in his rectangular backyard. If the backyard lot is 30 ft long by 35 ft wide, does he have enough fencing?

Since we are interested in surrounding the backyard with fencing, we are interested in the perimeter of the backyard. Its length is 30 ft, and its width is 35 ft, so $P = 2L + 2W = 2(30) + 2(35) = 60 + 70 = 130$ ft. Since, he only needs 130 ft of fencing material to surround his backyard, John will have enough fencing to erect his fence.

Answer: Yes, he has enough fencing.

Example 7

Juan has just had a pool added to his backyard. Due to the construction equipment and land clearing needed, all of his grass died. In order to sod the backyard, Juan needs to estimate how much sod he will need to order. His backyard is trapezoidal in shape with the dimensions shown below. Since his round pool has a diameter of 20 ft, how much sod will Juan need to order?

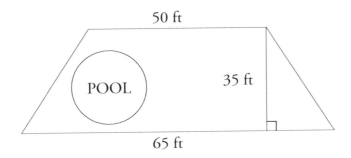

Since we know that he will not lay sod on top of or underneath the pool, we need to calculate the area of the backyard and subtract the area of the pool.

Area of Backyard – Area of Pool = Area of Sod Needed

Since the backyard is a trapezoid and the pool is a circle, the algebraic representation of the problem that needs to be solved follows.

$\frac{1}{2}h(a+b) - \pi r^2 = x$ $\qquad$ (x = area of sod needed.)

Substituting values for the variables above gives us the following.

$\frac{1}{2}(35)(50+65) - (3.14)(10)^2 = x$ $\qquad$ Remember the radius is $\frac{1}{2}$ the diameter. We will also use 3.14 as an approximation of pi.

$(.5)(35)(115) - (3.14)(100) = x$ $\qquad$ Use the order of operations to simplify.

$2012.5 - 314 = x$

$1698.5 = x$

Answer: Juan will need about $1,700$ ft² of sod to cover his backyard.

Note: Juan rounds up to a whole number just to be sure he has enough sod.

Simple Interest Word Problems

Let's review the simple interest formula from the previous section:
$I = Prt$

In words, this formula says that interest = (principal)(rate)(time), where the principal is the amount of money you start with, the rate is the interest rate paid by you or the bank, and time is measured in years.

Example 8

Marsha invested a certain amount of money at 4% and a different amount at 7%. The amount she invested at the lower rate was three times that of the amount at 7%. If the total interest accrued for both accounts in 1 year was $95.00, then how much did she invest in each account?

Here, money is being invested at 4% and at 7%. The total amount of interest earned is the sum of the interest earned at 4% and the interest earned at 7%.

We also know that "the amount she invested at the lower rate was three times the amount at 7%." In other words, (amount at 4%) = 3 (amount at 7%). Calling x the amount at 7%, amount at 4% = $3x$.

$I = Prt + Prt$ $\qquad$ Total interest is the sum of the two investments.

$I = P(.04)t + P(.07)t$ $\qquad$ One amount was invested at 4%, and the other was invested at 7%.

$95 = (3x)(.04)(1) + (x)(.07)(1)$ $\qquad$ Interest = 95, and time = 1 year.

$95 = .12x + .07x$ Multiply.

$95 = .19x$ Add.

So, $\dfrac{95}{.19} = \dfrac{.19}{.19}x$ Divide both sides by .19.

$500 = x$

x = the amount invested at 7%, so the amount invested at 4% is $3 \cdot 500 = 1,500$.

Answer: $1,500 was invested at 4%, and $500 was invested at 7%.

Check: $(1,500)(.04)(1) + (500)(.07)(1) = 60 + 35 = 95$.

Word Problems Involving Commission

In these problems, someone makes a certain amount of money based on how much s/he sells. The commission will be at a certain rate given in the problem. The formula is $C = PR$, where C = the amount of the commission, P = the amount of money paid for purchase of the item, and R = the commission rate. Let's look at some examples.

Example 9

If a computer sales representative earns 5.5% on her/his weekly sales, then how much will s/he earn at the end of a week in which s/he sold $5,200 worth of computers?

$C = PR$

$C = (\$5,200)(.055)$ Sales = $5,200, and commission rate = 5.5%.

Answer: $C = \$286$

Example 10

Jerry is trying to sell his house. In order to place the proper deposit on his next house, he will need to keep at least $60,000 from the sale of his current house, on which he owes $140,000. Knowing he will lose 8% of the selling price due to the agent's commission (and closing costs), what is the minimum amount he should sell his house for?

If Jerry wants to keep $60,000 of the money he gets for selling his house, he will certainly have to sell his house for more than $60,000 above what he owes on the house. Since he owes $140,000 and he wants $60,000, add the two numbers together for a total of $200,000. Remember that he has to pay the agent a commission as well as pay closing costs; thus, he will have to add some more to the price of his house. Let's call this extra amount of money x. Thus, the sales price = $200,000 + x$.

$C = PR$

$x = (200,000 + x)(.08)$ Sales price = $200,000 + x$, and commission = x.

$x = 16,000 + .08x$ Use the distributive property.

$x - .08x = 16,000 + .08x - .08x$ We need to subtract $.08x$ from both sides.

$.92x = 16,000$ Subtract $(1 - .08 = .92)$.

Thus, $\dfrac{.92x}{.92} = \dfrac{16,000}{.92}$ Divide both sides by .92.

$x = \$17,391.304$, which we can round down to $\$17,391$.

$\$200,000 + \$17,391 = \$217,391$

Answer: Jerry should sell his house for at least $\$217,391$.

Check: Commission will be $\$217,391 \cdot .08 \approx \$17,391$. Subtract the commission and what he owes for the house from the selling price: $\$217,391 - \$17,391 - \$140,000 = \$60,000$. Jerry still has enough money for his down payment.

Practice Problems

Translate into algebraic expressions. Do not solve the equations.

1. The sum of a number and four is six.
2. The difference between twice a number and four is the same as ten.
3. The product of seven and a number is equal to twelve.
4. The quotient of ten and a number is thirty.
5. Eight less than three times a number is nine.
6. One-fourth of a number is the same as the sum of the number and two.
7. Five more than a number is equal to five less than twice the number.
8. Six times the sum of a number and one is four.
9. Seven times the quotient of a number and three is six.
10. Half of the difference of a number and one is equal to fifteen.

Solve each of the following.

11. The sum of two consecutive integers is 45.
12. The sum of three consecutive integers is 54.
13. The sum of two consecutive odd integers is 92.
14. The sum of three consecutive odd integers is 33.
15. The sum of three consecutive even integers is 174.
16. The sum of two consecutive even integers is 74.
17. If the area of a triangle is 40 cm^2, find the base if the height is 5 cm.
18. If the area of a parallelogram is 120 cm^2, find the height if the base is 30 cm.
19. Find the circumference and area of a circle that has a diameter of 16 in.

20. Find the perimeter and area of a rectangle that has a base of 10 cm and a width of 4 cm.

21. A farmer has a large field that is rectangular in shape with a width of 1.5 miles and a length of 2 miles. If he wants to enclose the field with a chain link fence, how many miles of fencing will he need?

22. Using the information from problem 21, how many bags of fertilizer would he need if one bag covers $.03 \text{ mi}^2$?

23. Assume you have a concrete rectangular patio that measures $10 \text{ ft} \times 20 \text{ ft}$. A contractor has just added an in-ground Jacuzzi that measures 7 ft in diameter into the patio. If you want to cover the rest of the patio in ceramic tile, how much would you need?

24. Assume you have a wall in your bedroom that measures 15 ft long by 10 ft high. On this particular wall, there are two windows, each measuring 3 ft by 5 ft. If you want to wallpaper the wall, how much would you need to buy?

25. Lakesha invested a certain amount of money at 5% and an additional amount, which was double her investment at 5%, at 3%. If the total amount of interest accrued in both accounts was $330.00 at the end of one year, then how much did Lakesha invest in each account?

26. Mark has two simple interest loans in his name. In one, he borrowed $2,500 at 6%, and in the other, he borrowed $5,000 at 7.5%. If he has accrued $1,575 in interest, for how many years has he had the loans, assuming no payments have been made?

27. If a realtor earns 4.2% on the sale of a house, how much will the realtor's commission be on the sale of a house for $175,000?

28. Marcus earns commissions selling furniture on a tiered system. He receives 3% on the first $1,000 of furniture he sells; 4% on any additional amount up to $3,000; and 5% on everything above $3,000. How much of a commission would he receive in one week for sales totaling $9,000?

29. Determine what a seller would receive on a house that s/he sold for $290,000 if the following were applied: the seller owed $212,000 on the mortgage, the realtor receives 6% of the sales price, and the closing costs are around 3% of the sales price.

30. Bill is trying to sell his house. In order to place the proper deposit on his next house, he will need to get at least $40,000 from the sale of his current house, on which he owes $180,000. Knowing he will lose 9% of the selling price due to the agent's commission (and closing costs), what is the minimum amount he should sell his house for?

$x > 5$ looks like this:

$x < 1$ and $x > 5$ looks like this:

Answer: $\emptyset$

Example 13

Solve: $x < 1$ or $x > 5$

We are looking for numbers that are either smaller than 1 or bigger than 5 (or both, but there are no numbers that are both smaller than 1 and bigger than 5). Thus, the number line contains two different sets of numbers.

$x < 1$ looks like this:

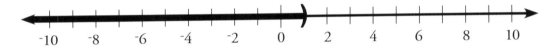

$x > 5$ looks like this:

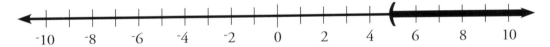

$x < 1$ or $x > 5$ looks like this:

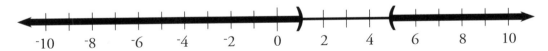

$x < 1$ or $x > 5$

Answer: $(-\infty, 1) \cup (5, \infty)$ Remember that $\cup$ stands for the union of both sets.

Example 14

Solve: $x - 3 \geq 4$ *or* $2x + 4 \leq 6$

$x - 3 \geq 4$ *or* $2x + 4 \leq 6$ Solve both inequalities. Add 3 to both sides on the left. Subtract 4 from both sides on the right.

$x \geq 7$ or $2x \leq 2$ Divide both sides on the right by 2.

$x \geq 7$ or $x \leq 1$

Answer: $(-\infty, 1] \cup [7, \infty)$ (Answer in interval notation)

Example 15

Solve: $x - 2 > 1$ and $x - 2 \leq 4$

Compound inequalities involving *and* are often simply written as a single inequality.

$1 < x - 2 \leq 4$

$1 + 2 < x - 2 + 2 \leq 4 + 2$ Add 2 to all three parts of the equation. Your goal is to isolate the x in the middle.

$3 < x \leq 6$

Answer: $(3, 6]$ This answer means that all numbers greater than 3 but less than or equal to 6 will satisfy the equation.

Example 16

Solve: $-2 < -2x + 6 < 0$

$-2 < -2x + 6 < 0$

$-2 - 6 < -2x + 6 - 6 < 0 - 6$ Subtract 6 from all three parts of the equation.

$\frac{-8}{-2} < \frac{-2x}{-2} < \frac{-6}{-2}$ Divide each term by –2. Since –2 is negative, the inequality signs change direction.

$4 > x > 3$ or $3 < x < 4$

Answer: $(3, 4)$

Note: Some teachers will prefer that you rewrite $4 > x > 3$ as $3 < x < 4$. They prefer the latter equation because it is traditional in inequalities to have the smaller number on the left and the larger number on the right.

Example 17

Solve: $\frac{2}{3} < \frac{x + 1}{4} \leq 2$

$\frac{2}{3} < \frac{x+1}{4} \leq 2$ The common multiple is 12.

$12\left(\frac{2}{3}\right) < 12\left(\frac{x+1}{4}\right) \leq 12(2)$ Multiply everywhere by 12.

$8 < 3(x + 1) \leq 24$ Use the distributive property.

$8 < 3x + 3 \leq 24$ Subtract 3 from all parts of the equation.

$8 - 3 < 3x + 3 - 3 \leq 24 - 3$

$5 < 3x \leq 21$ Divide all parts by 3.

$\frac{5}{3} < \frac{3x}{3} \leq \frac{21}{3}$

$\frac{5}{3} < x \leq 7$

Answer: $\left(\frac{5}{3}, 7\right]$

Practice Problems

Solve the following inequalities and graph the solutions.

1. $x - 3 > 7$
2. $x + 8 < 2$
3. $x - 6 \leq -1$
4. $x + 2 \geq -4$
5. $3x > 12$
6. $-4x < 36$
7. $-6x \leq -10$
8. $4x \geq -14$
9. $3x - 7 > 5$
10. $-7x - 21 < 49$

Solve the following inequalities and write the answers using interval notation.

11. $-2x + 5 \geq 5 + x$
12. $6x - 4 \leq 13 - 5x$
13. $8 - 2x \leq 5$
14. $3 - x > 0$
15. $3(x - 4) > 2x - 7$
16. $-4(2x + 5) - 1 \leq 5 - 3x$
17. $\frac{1}{2}x - \frac{4}{3} > 3$

18. $\frac{-2}{5}x + \frac{3}{4} \leq \frac{5}{2}$

19. $\frac{x+4}{4} \geq 3$

20. $\frac{5-2x}{3} < \frac{1}{2}$

Solve the following compound inequalities and write the solutions using interval notation.

21. $4(2x-5) < 8x+1$
22. $-2(x-6) \geq 12 - 2x$
23. $x > 4$ and $x < 6$
24. $x > 0$ and $x < 4$
25. $x > -2$ and $x \leq -3$
26. $x \geq 4$ and $x < 2$
27. $x \geq 3$ and $x \geq 6$
28. $x \leq 0$ and $x \leq -2$
29. $x > -1$ or $x < 2$
30. $x \geq 4$ or $x \leq 6$
31. $x < -1$ or $x > 3$
32. $x < 3$ or $x > 0$
33. $x < -7$ or $x \leq -2$
34. $x \geq 0$ or $x > 5$
35. $x - 3 > 5$ or $x + 1 < 4$
36. $2x - 4 \geq 6$ or $3x \leq 12$
37. $5x - 10 > 5$ and $8x - 4 < 4$
38. $5x - 10 < 5$ and $8x - 4 > 4$
39. $0 \leq 3x - 9 < 6$

Solve the following compound inequalities and graph the solutions.

40. $-2 < 2x + 10 < 8$
41. $-4 \leq -4x + 2 \leq 6$
42. $-1 < 5 - x < 4$
43. $-3 \leq \frac{1}{2}x - 6 < 0$
44. $\frac{-2}{3} < \frac{1}{4}x - 2 < 1$

45. $-1 \leq \dfrac{x+4}{4} \leq 2$

46. $-3 < \dfrac{6-x}{5} < 3$

Section 5
Absolute Value Equations and Inequalities

Learning Objectives

When you finish your study of this section, you should be able to
- Solve equations involving absolute value signs
- Solve inequalities involving absolute value signs

Solving Absolute Value Equations

Recall that the absolute value of a number is its distance from 0. But, when measuring the distance from 0, we can go in two different directions: the positive one or the negative one. For this reason, an absolute value equation is really two equations in disguise.

To solve $|something| = q$, (as long as q is not 0), solve these two equations: $something = q$ and $something = -q$.

Let's look at some examples.

Example 1

Solve: $|x| = 6$

Set the expression inside the absolute value signs equal to 6 and -6.

Answer: $x = 6$ or $x = -6$

Example 2

Solve: $|x - 3| = 4$

Set the expression inside the absolute value signs equal to 4 and -4.

$x - 3 = 4$ or $x - 3 = -4$ Add 3 to both sides.

Answer: $x = 7$ or $x = -1$

Example 3

Solve: $2|2x - 7| = 8$

We cannot rewrite this equation until we have the absolute value sign **by itself** on the left side.

$2|2x - 7| = 8$ Divide both sides by 2.

Chapter 2: Solving Various Equations

$\|2x-7\|=4$	Rewrite.
$2x-7=4$ or $2x-7=-4$	Add 7 to both sides.
$2x=11$ or $2x=3$	Divide both sides by 2.

Answer: $x=\dfrac{11}{2}$ or $x=\dfrac{3}{2}$

Example 4

Solve: $\|3x-2\|=-4$

Recall that the absolute value of any number is positive or 0. It cannot be negative. Thus, this equation has no solution. Don't even try to rewrite it.

Answer: no solution

Solving Absolute Value Inequalities Involving *Less Than*

These inequalities can be rewritten into a single inequality, one that involves *and*.
- To solve $|\,something\,| < q$ (as long a q is not 0), solve $-q < something < q$.
- To solve $|\,something\,| \leq q$ (as long a q is not 0), solve $-q \leq something \leq q$.

Let's look at some examples.

Example 5

Solve: $|x| < 4$

Rewrite as $-4 < x < 4$.

Answer: $-4 < x < 4$ or, in interval notation, $(-4, 4)$

Example 6

Solve: $|3x-6| \leq 9$

$-9 \leq 3x - 6 \leq 9$	Rewrite.
$-9 + 6 \leq 3x - 6 + 6 \leq 9 + 6$	Add 6 to all parts of the equation.
$-3 \leq 3x \leq 15$	Divide each term by 3.

Answer: $-1 \leq x \leq 5$ or, in interval notation, $[-1, 5]$

Example 7

Solve: $|2x + 1| < -4$

Recall that the absolute value of any number is positive or 0. The absolute value cannot be negative, so it certainly can't be smaller than -4. This inequality has no solution. Don't even try to rewrite it.

Answer: no solution or $\emptyset$

Example 8

Solve: $3|x + 5| + 6 \leq 12$

We cannot rewrite the inequality until the absolute value is by itself on the left.

$3\|x + 5\| + 6 \leq 12$	Subtract 6 from both sides.
$3\|x + 5\| \leq 6$	Divide both sides by 3.
$\|x + 5\| \leq 2$	Now, rewrite.
$-2 \leq x + 5 \leq 2$	The inequality is in rewritten form.
$-2 - 5 \leq x + 5 - 5 \leq 2 - 5$	Subtract 5 from all parts of the inequality.

Answer: $-7 \leq x \leq -3$ or, in interval notation, $[-7, -3]$

Example 9

Solve: $|6 - 2x| < 8$

$-8 < 6 - 2x < 8$	The inequality is in rewritten form.
$-8 - 6 < 6 - 6 - 2x < 8 - 6$	Subtract 6 from all parts of the inequality.
$-14 < -2x < 2$	Divide all parts of the inequality by -2. Since -2 is negative, the inequality signs change direction.
$7 > x > -1$	

Answer: $-1 < x < 7$ or, in interval notation, $(-1, 7)$ Rewrite with smaller number on the left.

Solving Absolute Value Inequalities Involving *Greater Than*

These inequalities can be rewritten into two inequalities involving *or*.
- To solve $|something| > q$ (as long a q is not 0), solve *something* $> q$ or *something* $< -q$.
- To solve $|something| \geq q$ (as long a q is not 0), solve *something* $\geq q$ or *something* $\leq -q$.

Let's look at some examples.

Example 10

Solve: $|x| > 4$

Rewrite as $x > 4$ or $x < -4$.

Answer: $x > 4$ or $x < -4$

or, in interval notation,

$(-\infty, -4) \cup (4, \infty)$.

Note: Recall that you combine two sets using the union symbol.

Example 11

Solve: $|x + 6| - 1 \geq 4$

Add 1 to both sides to isolate the | | sign on the left side.

$|x + 6| \geq 5$	Rewrite.
$x + 6 \geq 5$ or $x + 6 \leq -5$	Subtract 6 from both sides of each equation.

Answer: $x \geq -1$ or $x \leq -11$

or, in interval notation, $(-\infty, -11] \cup [-1, \infty)$.

Example 12

Solve: $5|5x - 10| \geq 10$

First, divide both sides by 5 so that the absolute value part is by itself on the left side.

$|5x - 10| \geq 2$	Rewrite.
$5x - 10 \geq 2$ or $5x - 10 \leq -2$	Add 10 to both sides of each equation.
$5x \geq 12$ or $5x \leq 8$	Divide both sides by 5.

Answer: $x \geq \dfrac{12}{5}$ or $x \leq \dfrac{8}{5}$

or, in interval notation, $\left(-\infty, \dfrac{8}{5}\right] \cup \left[\dfrac{12}{5}, \infty\right)$

Example 13

Solve: $|2x - 7| > -2$

This inequality is unusual. The absolute value of any number is positive or 0. Thus, in particular, the absolute value will always be greater than any negative number, including -2.

Answer: all real numbers or, in interval notation, $(-\infty, \infty)$

Example 14

Solve: $|2x - 6| \leq 0$

This inequality is also unusual. The absolute value of any number is always greater than or equal to zero. Most students think the answer is that no solution exists, but there is an equal sign in the problem. The equal sign means there is one solution, the value of x that makes the expression inside the absolute value signs equal to zero.

$2x - 6 = 0$	Add 6 to both sides.
$2x = 6$	Divide both sides by 2.
Answer: $x = 3$	3 is the only answer. (Anything else creates a value that is greater than 0.)

SUMMARY OF ABSOLUTE VALUE PROBLEMS

Original	Rewrite As
$|\text{something}| = q$	something $= q$ or something $= -q$
$|\text{something}| < q$	$-q <$ something $< q$
$|\text{something}| > q$	something $> q$ or something $< -q$

Practice Problems

Solve the following equations.

1. $|x| = 2$
2. $|x| = -3$
3. $|x - 5| = 9$
4. $|x - 4| = 3$
5. $|3 - x| = 6$
6. $|8 - 2x| = 4$
7. $|3x - 1| = 6$
8. $|4x - 2| - 3 = 5$
9. $|3x - 5| + 2 = 6$
10. $|5x + 2| = 5$
11. $|5x - 1| = -3$
12. $|6x - 5| = -6$

13. $|8x-1|=0$
14. $|3x+9|=0$
15. $5|x-2|=15$
16. $2|x+1|=10$
17. $-4|x+3|-5=-9$
18. $3|2x+3|+5=3$

Solve the following inequalities and write your answer using interval notation.

19. $|x|>3$
20. $|x|\geq 9$
21. $|x|\leq 4$
22. $|x|<1$
23. $|x+6|<6$
24. $|x-2|\leq 4$
25. $|x+3|>6$
26. $|x-9|\geq 1$
27. $|x-7|-2<4$
28. $|x+4|+3>8$
29. $|5-2x|>5$
30. $|6-3x|\leq 9$
31. $|2+3x|>-5$
32. $|7x-9|<0$
33. $2|2x+6|<18$
34. $4|3x-1|>16$
35. $3|2x+3|-3\geq 9$
36. $-4|4x-8|-4<8$
37. $5|3-2x|+1>6$
38. $6|3x-4|+3\leq 6$
39. $|3x-2|+4\leq 4$
40. $|8x-6|+2\geq 2$

Chapter 3

Functions and Graphing

ASSIGNMENT CHECKLIST

What You Should Do	Where?			When?	✓
Read, view the videos, and then complete the online work for Chapter 3, Section 1	📖	💻	MathXL	After completing Chapter 2	
Read, view the videos, and then complete the online work for Section 2	📖	💻	MathXL	After completing Chapter 3, Section 1	
Read, view the videos, and then complete the online work for Section 3	📖	💻	MathXL	After completing Section 2	
Read, view the videos, and then complete the online work for Section 4	📖	💻	MathXL	After completing Section 3	
Read, view the videos, and then complete the online work for Section 5	📖	💻	MathXL	After completing Section 4	
Take the quiz on Chapter 3			MathXL	After completing Section 5	
Take the practice test on Chapters 1-3			MathXL	After completing Chapter 3 quiz	
Schedule your test with your instructor				After completing practice test on Chapters 1-3	
Post questions and respond to other students' questions in the Discussion Board		💻		Anytime	
Other assignments:					
Notes:					

Section 1
Introduction to Functions and Graphing Lines

Learning Objectives

When you finish your study of this section, you should be able to
- Plot points on the coordinate axis and identify in which quadrant the point lies
- Graph a line given in slope-intercept form or standard form
- Graph a horizontal or vertical line
- Understand function notation

Rectangular Coordinate System

The **rectangular coordinate system** is divided into four **quadrants** (Quadrant I, Quadrant II, Quadrant III, and Quadrant IV). The x-axis is horizontal; the y-axis is vertical. A **point** is a pair of numbers written in parentheses, for example, (x,y). The first number is called the **x-coordinate**. The second number is called the **y-coordinate**.

Plotting Points In The Rectangular Coordinate System

Notice that the x-coordinate tells you how far to move from the point $(0,0)$ (also called the **origin**) horizontally. The y-coordinate tells you how far to move from the point $(0,0)$ vertically. Let's look at some examples.

Example 1

Plot the points $(-2,4)$, $(5,-1)$, and $(3,0)$.

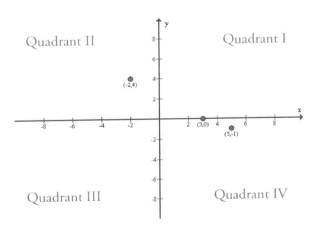

Example 2

In which quadrant does the point $(-9, 15)$ lie?

Answer: Since the x-coordinate, -9, is negative, and the y-coordinate, 15, is positive, this point must be in Quadrant II.

Example 3

In which quadrant does the point $(.0001, .0004)$ lie?

Answer: Since both coordinates are positive, this point must be in Quadrant I.

Example 4

In which quadrant does the point $(100, 0)$ lie?

This question is a trick. The point $(100, 0)$ is located on the x-axis. It is not in any of the four quadrants.

Example 5

In which quadrant does a point in the form of $(x, -y)$ lie? (Assume x and y both represent positive numbers.)

To answer this question, let's pick values for x and y; let $x = 3$ and $y = 5$. Then, $-y = -(5) = -5$. Thus, the point would be $(3, -5)$.

Answer: Since the x-coordinate is positive and the y-coordinate is negative, the point must be in Quadrant IV.

Graphing Linear Equations and Functions

GRAPHING LINEAR FUNCTIONS IN SLOPE-INTERCEPT FORM

To graph lines given in slope-intercept form, we create a table of y-values, given certain x-values. We recommend that you pick at least three x-values before trying to draw the line. You will see that we often pick more than three x-values. You can pick any numbers you want to substitute for x; try to pick numbers that make your arithmetic relatively easy.

Note: Sometimes, $y = mx + b$ is written in function notation: $f(x) = mx + b$. Just note that y and $f(x)$ represent the same thing. In Section 5 we will discuss function notation further.

Example 6

Graph: $f(x) = 4x - 3$

We will let $x = 2, 1, 0,$ and -1. For each chosen x-value, we calculate the corresponding y-value using the given equation. The pairs of numbers give us the points on the graph of the line $f(x) = 4x - 3$.

x	$f(x) = 4x - 3$	Point
2	$4(2) - 3 = 5$	$(2, 5)$
1	$4(1) - 3 = 1$	$(1, 1)$
0	$4(0) - 3 = -3$	$(0, -3)$
-1	$4(-1) - 3 = -7$	$(-1, -7)$

Now, we simply plot the four points and draw a smooth, straight line through them:

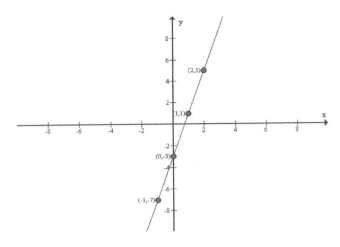

Example 7

Graph: $f(x) = \dfrac{-1}{3}x + 1$

Since we will have to multiply the x-values by $\dfrac{-1}{3}$, it would make sense to pick multiples of 3 for x-values. We pick $x = 0, 3,$ and 6.

x	$f(x) = \dfrac{-1}{3}x + 1$	Point
0	$-1/3(0) + 1 = 1$	$(0, 1)$
3	$-1/3(3) + 1 = 0$	$(3, 0)$
6	$-1/3(6) + 1 = -1$	$(6, -1)$

Plot the points and draw the line.

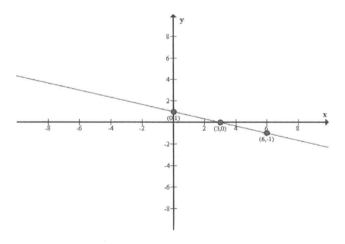

GRAPHING LINEAR EQUATIONS IN STANDARD FORM USING THE INTERCEPT METHOD

To graph lines given in standard form, we find the **x-intercept**, the point where the line crosses the x-axis, and the **y-intercept**, the point where the line crosses the y-axis.

To find the x-intercept, let $y = 0$ and solve for x.
To find the y-intercept, let $x = 0$ and solve for y.
Let's look at some examples.

Example 8

Graph: $2x + y = 4$

Find the x-intercept:

If $y = 0$, then by substitution the new equation is $2x + 0 = 4$, so

$2x = 4$ Divide both sides of the equation by 2.

$x = 2$.

Thus, when $x = 2$, $y = 0$; that is, the line passes through the point $(2, 0)$.

Find the y-intercept:

If $x = 0$, then by substitution the new equation is $2(0) + y = 4$, so

$y = 4$.

Thus, when $x = 0$, $y = 4$; that is, the line passes through the point $(0, 4)$.

Plot these two points and draw a smooth, straight line through them.

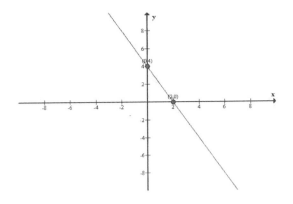

Example 9

Graph: $5x - 2y = 5$

Find the x-intercept:

If $y = 0$, then by substitution the new equation is $5x - 2(0) = 5$, so

$5x = 5$

$x = 1$.

Thus, when $x = 1$, $y = 0$; that is, the line passes through the point $(1, 0)$.

Find the y-intercept:

If $x = 0$, then by substitution the new equation is $5(0) - 2y = 5$, so

$-2y = 5$

$\dfrac{-2y}{-2} = \dfrac{5}{-2}$

$y = -2.5$.

Thus, when $x = 0$, $y = -2.5$; that is, the line passes through the point $(0, -2.5)$.

Now, plot the two points and draw a smooth, straight line through them.

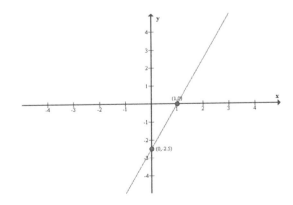

Example 10

Graph: $x - 2y = 0$

Find the x-intercept:

If $y = 0$, then by substitution the new equation is $x - 2(0) = 0$, so

$x = 0$.

The x-intercept is $(0,0)$.

Find the y-intercept:

If $x = 0$, then by substitution the new equation is $0 - 2y = 0$, so

$-2y = 0$.

Dividing both sides by -2 gives us the following:

$y = 0$.

The y-intercept is also $(0,0)$.

Now we are in a bind. To graph a line we need **two distinct points**. So far, we have only the point $(0,0)$. We need to pick some other value of x and solve for y. Let's let $x = 2$. Then,

$2 - 2y = 0$

$2 - 2 - 2y = 0 - 2$

$-2y = -2$

$y = 1$.

Thus, when $x = 2$, $y = 1$; that is, the line passes through the point $(2,1)$.

Plot the two points and draw the line.

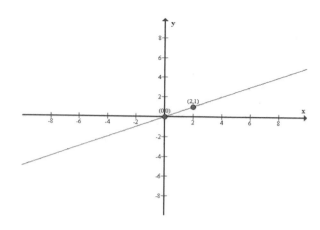

Chapter 3: Functions and Graphing

Graphing Horizontal and Vertical Lines

A **horizontal line** is $f(x) = c$ (or $y = c$), where c is any number.

A **vertical line** is $x = c$, where c is any number.

These definitions may seem backwards to you, so it's worth spending some time explaining why horizontal and vertical lines are defined the way they are.

Example 11

Graph: $x = -2$

This line says that the x-coordinate of points on this line is always -2, but the y-coordinate can be any number we like. For example, the points $(-2, -2)$, $(-2, 0)$, and $(-2, 2)$ are on this line. If you plot these three points, you will see that they do fall on a vertical line. That is the reason we call this vertical line $x = -2$; all the points that lie on it have an x-coordinate of -2.

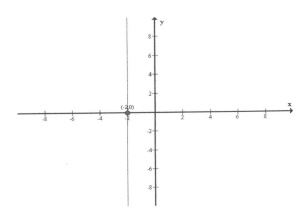

Example 12

Graph: $y = 4$

This line says that the y-coordinate of all of the points on this line is 4, but x can have any value we like. For example, the points $(-2, 4)$, $(0, 4)$, and $(2, 4)$ are on this line. If you plot these three points, you will see that they do indeed lie on a horizontal line.

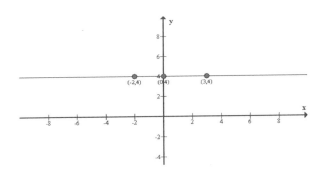

GRAPHING LINES WITH LARGE SCALES

As a final example, let's do this problem:

Example 13

Graph: $f(x) = 5x + 120$

We will create a table, letting $x = -1$ and 1.

x	$f(x) = 5x + 120$	Point
-1	$5(-1) + 120 = 115$	$(-1, 115)$
1	$5(1) + 120 = 125$	$(1, 125)$

If you start counting on your y-axis at 1, you will be drawing a lot of tick marks to get up to 115 and 125. Since these numbers are so big, we count on the y-axis by a larger number. Let's count by $20s: -60, -40, -20, 0, 20, 40, 60 \ldots$

We show the graph below with the two points labeled on it.

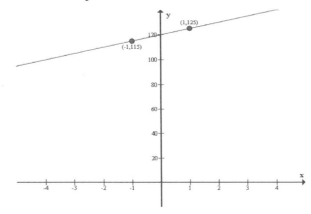

Practice Problems

In which quadrant or on what axis do the following points lie?

1. $(-4, 8)$
2. $(-1, -20)$
3. $(5, 9)$
4. $(100, -1)$
5. A point with the following characteristics: $(-x, -y)$ (assume positive x and y)
6. A point with the following characteristics: $(-x, y)$ (assume positive x and y)

7. $(20, 0)$
8. $(0, -30)$
9. $(-.0002, -.06)$
10. $(.0056, -.002)$

Plot the following points in the coordinate plane.

11. $(-2, 1)$
12. $(4, 3)$
13. $(-5, -2)$
14. $(4, -3)$
15. $(0, 5)$
16. $(6, 0)$
17. $\left(\frac{3}{2}, -2\right)$
18. $\left(5, \frac{5}{2}\right)$

Find the $x-$ and $y-$ intercepts of the following equations. Do not graph.

19. $f(x) = 2x - 1$
20. $f(x) = 5x + 4$
21. $f(x) = \frac{1}{3}x - 5$
22. $f(x) = \frac{-3}{2}x - 1$
23. $2x + 3y = 6$
24. $3x - 6y = -12$
25. $x + 3y = 9$
26. $x - y = 4$
27. $5x - 6y = 10$
28. $3x + 2y = 5$
29. $x - y = 0$
30. $3x - y = 0$

31. $\frac{1}{4}x - \frac{1}{2}y = -2$

32. $\frac{2}{3}x + \frac{1}{5}y = 3$

Graph the following equations on the coordinate plane. You may use either method.

33. $f(x) = 3x - 4$
34. $f(x) = 2x + 5$
35. $f(x) = -x + 1$
36. $f(x) = -5x + 3$
37. $3x + y = 6$
38. $4x - 2y = 8$
39. $x + 5y = -5$
40. $x - 3y = 9$
41. $4x - 3y = 8$
42. $3x + 5y = -10$
43. $4x - 2y = -5$
44. $6x + 4y = 10$
45. $2x + y = 0$
46. $x + y = 0$
47. $x = -4$
48. $y = 1$
49. $y = -5$
50. $x = 3$
51. $f(x) = \frac{-1}{3}x + 4$
52. $f(x) = \frac{1}{2}x - 2$
53. $f(x) = \frac{-3}{5}x - 3$
54. $f(x) = \frac{3}{4}x + 2$
55. $\frac{1}{2}x - \frac{1}{3}y = -2$
56. $\frac{5}{3}x + \frac{1}{5}y = 1$
57. $f(x) = 3x - 60$
58. $f(x) = -4x + 100$

Section 2
The Slope of a Line

Learning Objectives

When you finish your study of this section, you should be able to
- Find the slope of a line when given two points or given a graph
- Find the slope when given the equation of a line
- Determine whether two lines are parallel, perpendicular, or neither
- Determine if given points determine certain geometric shapes

Finding the Slope of a Line

Finding the Slope, Given Two Points

The **slope** of a line measures how "steep" it is numerically. Given two points, (x_1, y_1) and (x_2, y_2), the slope, which is usually represented by the letter m, can be calculated using the formula

$$m = \frac{y_2 - y_1}{x_2 - x_1}.$$

Slopes can be any real number or even undefined. Let's look at some examples.

Example 1

Find the slope of the line that passes through the points $(2, 3)$ and $(-1, 4)$.

It doesn't matter which point we call the "first point" and which point we call the "second point." Let's use the two points in the order they were given. Thus, $x_1 = 2, y_1 = 3$, and $x_2 = -1, y_2 = 4$

$$m = \frac{y_2 - y_1}{x_2 - x_1}$$

$$m = \frac{(4) - (3)}{(-1) - (2)}$$

Answer: $m = \frac{1}{-3}$

Example 2

Find the slope of the line that passes through the points $(-6, 0)$ and $(-2, -5)$.

$$m = \frac{(-5)-(0)}{(-2)-(-6)}$$

Answer: $m = \dfrac{-5}{4}$

Example 3

Find the slope of the line that passes through the points $(1, 4)$ and $(1, -3)$.

$$m = \frac{(-3)-(4)}{(1)-(1)}$$

Answer: $m = \dfrac{-7}{0}$ = undefined

Note: Recall that we cannot divide by zero; such an expression is referred to as "undefined." A line whose slope is undefined is vertical. A line with slope equal to 0 is horizontal. A handy chart follows:

Type of Line	Equation	Slope
Vertical	$x = c$	Undefined
Horizontal	$y = c$	0

FINDING THE SLOPE, GIVEN THE GRAPH OF A LINE

If you are given the graph of a straight line, we recommend finding the x-intercept and y-intercept and then using those two points to find the slope. Let's look at some examples.

Example 4

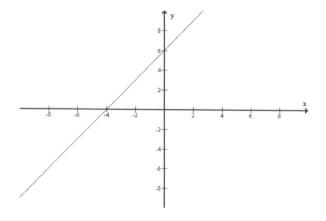

The x-intercept is $(-4, 0)$. The y-intercept is $(0, 6)$.

Thus, $m = \dfrac{0 - 6}{-4 - 0} = \dfrac{-6}{-4} = \dfrac{3}{2} = 1.5$.

Example 5

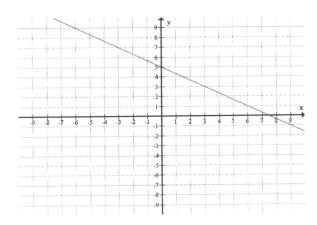

This problem is a bit tricky. The x-intercept is a decimal, so we just can't assume what decimal it actually is. The y-intercept is $(0, 5)$. We still need one more point. If you look over the graph, you will see some other points we could use, even though they are not intercepts, such as $(-3, 7)$, $(3, 3)$, $(6, 1)$, etc. We will use $(6, 1)$ as the second point to find the slope.

Thus, $m = \dfrac{5 - 1}{0 - (6)} = \dfrac{4}{-6}$ or $\dfrac{-2}{3}$.

Example 6

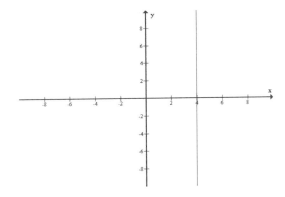

This line is vertical; therefore, its slope is undefined. No calculations are necessary.

Answer: undefined

FINDING THE SLOPE, GIVEN THE EQUATION OF A LINE OR LINEAR FUNCTION

If a line is given to you in slope-intercept form, $f(x) = mx + b$ (or $y = mx + b$), the slope is m. If the line is not in slope-intercept form, change it to that form before trying to find the slope. Let's look at some examples.

Example 7

Find the slope: $y = \dfrac{-2}{3}x + 5$

Answer: This line is already in slope-intercept form. The slope is $-2/3$.

Example 8

Find the slope: $5x + y = 7$

This line is not in slope-intercept form, so first we change it to that form.

$5x + y = 7$	To isolate y, subtract $5x$ from both sides.
$5x - 5x + y = -5x + 7$	Subtract.
$y = -5x + 7$	

Answer: The slope is -5.

Example 9

Find the slope: $3x - 2y = 9$

First, change the equation to slope-intercept form.

$3x - 2y = 9$	To isolate $-2y$, subtract $3x$ from both sides.
$3x - 3x - 2y = -3x + 9$	Subtract.
$-2y = -3x + 9$	
$\dfrac{-2y}{-2} = \dfrac{-3x}{-2} + \dfrac{9}{-2}$	Divide both sides by -2.
$y = \dfrac{3}{2}x - \dfrac{9}{2}$	

Answer: The slope is $3/2$.

Optional: If a line is given to you in standard form, $Ax + By = C$, the slope is $m = \dfrac{-A}{B}$

Thus, in Example 9 above, we could have just calculated that $m = \dfrac{-(3)}{-2} = \dfrac{3}{2}$

Determining Whether Two Lines Are Parallel, Perpendicular, or Neither

Two lines are **parallel** if their slopes are equal.

Two lines are **perpendicular** if the product of their slopes is -1. In other words, if you multiply the two slopes together, the product is -1 if the lines are perpendicular. Two such slopes are called **negative reciprocals**.

Let's look at some examples.

Example 10

Classify as parallel, perpendicular, or neither: $5x + y = 9$ and $-2x + 10y = 3$

(We will use the shortcut noted at the end of Example 9.)

The slope, m, of $5x + 1y = 9$ is $\frac{-5}{1} = -5$.

The slope, m, of $-2x + 10y = 3$ is $\frac{-(-2)}{10} = \frac{1}{5}$.

Now, the two slopes are certainly not equal, so the lines are not parallel.

If we multiply the two slopes together, however, the product is $-5 \cdot \frac{1}{5} = -1$.

Answer: Since the product of the slopes is -1, the two lines are **perpendicular**.

Example 11

Classify as parallel, perpendicular, or neither: $6x - 4y = 3$ and $2x + 5y = 6$

The slope, m, of $6x - 4y = 3$ is $m = \frac{-(6)}{-4} = 1.5$.

The slope, m, of $2x + 5y = 6$ is $m = \frac{-2}{5} = -0.4$.

The two slopes are not equal, so the lines aren't parallel.

If we multiply the slopes together, $1.5 \cdot (-0.4)$, their product is -0.6, so, the lines aren't perpendicular, either. Thus, the lines are neither parallel nor perpendicular.

Answer: neither

Example 12

Classify as parallel, perpendicular, or neither: $7x - 2y = 14$ and $14x - 4y = 12$

The slope, m, of $7x - 2y = 14$ is $m = \frac{-(7)}{-2} = 3.5$.

The slope, m, of $14x - 4y = 12$ is $\frac{-14}{-4} = 3.5$.

The slopes are equal, so the lines are parallel.

Answer: parallel

Determining Whether Given Points Determine a Geometric Shape

Two geometric shapes will be studied in this section: right triangles and parallelograms. By definition, a **right triangle** is a triangle containing exactly one right angle. Two lines that form a right angle are, by definition, perpendicular.

A **parallelogram** is simply a four-sided polygon whose opposite sides are parallel.

Example 13

Determine if the following points form a right triangle: $(0,4)$, $(1,1)$, and $(3,5)$

If the three points given form a right triangle, two of the three slopes, when multiplied together, will equal -1.

Let's calculate the slopes of each side of the right triangle..

Find the slope of the straight line connecting $(0,4)$ to $(1,1)$:

$$m = \frac{4-1}{0-1} = -3 \text{ or } \frac{-3}{1}$$

Find the slope of the straight line connecting $(0,4)$ to $(3,5)$:

$$m = \frac{4-5}{0-3} = \frac{1}{3}$$

Find the slope of the straight line connecting $(1,1)$ to $(3,5)$:

$$m = \frac{1-5}{1-3} = \frac{-4}{-2} = \frac{2}{1} = 2$$

Are there two slopes whose product is -1? Yes, $(-3) \cdot \left(\frac{1}{3}\right) = -1$

Thus, those two lines are perpendicular. Therefore, the three sides do contain a right angle.

Answer: Yes, the points form a right triangle.

Example 14

Determine if the following points form a parallelogram: $(0,2)$, $(2,5)$, $(4,1)$, and $(6,6)$

If the four points given form a parallelogram, opposite sides will be parallel. This means if we calculate the slopes of all four sides, the slopes of the opposite sides should be the same.

Let's calculate the slope of each side of the possible parallelogram.

Find the slope of the straight line connecting $(0,2)$ and $(2,5)$.

$$m = \frac{5-2}{2-0} = \frac{3}{2}$$

Find the slope of the straight line connecting $(4,1)$ and $(6,6)$.

$$m = \frac{6-1}{6-4} = \frac{5}{2}$$

Find the slope of the straight line connecting $(2,5)$ and $(6,6)$.

$$m = \frac{6-5}{6-2} = \frac{1}{4}$$

Finally, find the slope of the straight line connecting $(0,2)$ and $(4,1)$.

$$m = \frac{1-2}{4-0} = -\frac{1}{4}$$

Are any of the slopes equal? The answer is no. Thus, this shape is not a parallelogram. In geometric terms, these points simply create a quadrilateral, which is simply a four-sided polygon.

Word Problems Involving Slope

Example 15

The pitch of a roof is its slope. Find the pitch of the roof shown.

In this problem, it's helpful to think of the slope as $\frac{\text{rise}}{\text{run}}$. The "rise" of the roof, or its vertical distance, is 15 ft. Its "run," or horizontal distance, is 25 ft.

Thus, the pitch is

$$\frac{15}{25} = 0.6$$

Example 16

Driving down a mountain, Jerry finds he descends 2500 ft in elevation by the time he is 5 mi away (horizontally) from the high point on the mountain road. Find the slope of his descent. (Hint: 1 mi = 5,280 ft)

Since Jerry is descending, his "rise" is -2500 ft. The "run" is 5 mi.

Thus, slope $= \dfrac{\text{rise}}{\text{run}} = \dfrac{-2500 \text{ ft}}{5 \text{ mi}} \cdot \dfrac{1 \text{ mi}}{5280 \text{ ft}} = -.095$

Practice Problems

Find the slope of the line passing through the following pairs of points.

1. $(-2, 4), (5, 1)$
2. $(-4, 0), (6, -2)$
3. $(-2, -1), (0, 4)$
4. $(-5, -4), (4, 3)$
5. $(-3, 2), (2, 2)$
6. $(1, 2), (1, -6)$

Find the slope of the line shown in each graph.

7.

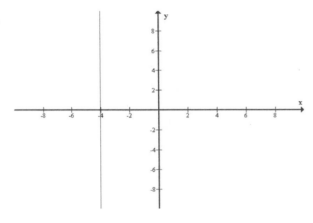

8.

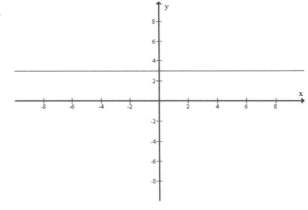

9.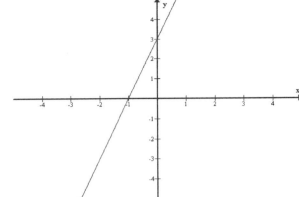

10.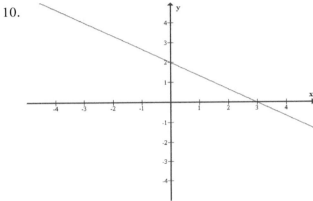

11.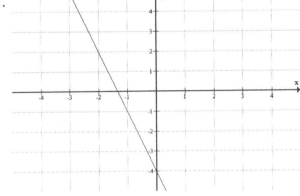

12.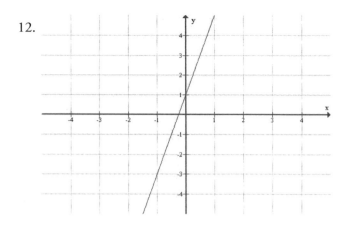

Find the slope of the line, given the following equations.

13. $f(x) = 4x - 5$
14. $f(x) = \frac{-2}{3}x + 4$
15. $3x + y = 6$
16. $7x + y = -4$
17. $5x - y = 2$
18. $-2x - y = 8$
19. $4x - 2y = 8$
20. $5x - 10y = 20$
21. $2x + 4y = 9$
22. $3x - 9y = 7$
23. $2x + 7y = 4$
24. $5x - 2y = 7$

Determine if the following lines are parallel, perpendicular, or neither.

25. $2x - y = 4; 3x + y = 5$
26. $5x + y = 10; 10x + 2y = -2$
27. $4x + 2y = 8; 6x - 3y = 5$
28. $x + 3y = 7; 3x - 2y = -5$

Determine if the following points form a right triangle.

29. $(-1, 4), (2, 0), (3, 5)$
30. $(-5, 3), (1, 5), (3, -1)$

Determine if the following points create a parallelogram.

31. $(-1, 0), (-3, 4), (2, 0), (0, 4)$

32. $(-1, -1), (0, 4), (3, -2), (5, 1)$

The pitch of a roof is its slope. Find the pitch of the roof shown.

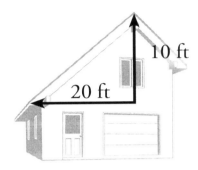

33. Driving down a mountain, Laura finds she descends $2,200$ ft in elevation by the time she is 4.8 mi away (horizontally) from the high point on the mountain road. Find the slope of her descent. (Hint: $1 \text{ mi} = 5,280 \text{ ft}$)

Section 3
Finding Equations of Lines

Learning Objectives

When you finish your study of this section, you should be able to
- Find the equation of a line, given its graph
- Use point-slope form to find the equation of a line
- Find the equation of a line, given two points
- Find the equations of horizontal and vertical lines
- Writing equations in standard and slope-intercept form.

Linear Equations

Linear equations, equations that represent straight lines, come in several forms. In this section, we will look at slope-intercept form and standard form.

Slope-Intercept Form: $y = mx + b$, where m is called the slope and b is called the y-intercept.
Standard Form: $Ax + By = C$.

Example 1

Write the following equation in slope intercept form: $4x - 2y = 10$

To change an equation from standard form to slope-intercept form, we solve for y.

$4x - 2y = 10$	To isolate y, subtract $4x$ from both sides.
$4x - 4x - 2y = -4x + 10$	Subtract.
$-2y = -4x + 10$	
$\dfrac{-2y}{-2} = \dfrac{-4x}{-2} + \dfrac{10}{-2}$	Divide each term by -2.

Answer: $y = 2x - 5$

Example 2

Write the following equation in standard form: $2x - 5y - 11 = -3$

$2x - 5y - 11 = -3$	We need to move -11.

$2x - 5y - 11 + 11 = -3 + 11$ Add 11 to both sides.

Answer: $2x - 5y = 8$.

Example 3

Write the following equation in standard form: $2(x - 1) = -3(2y - 4)$

Our goal is to simplify and isolate the x and y terms together on the left side of the equation.

$2(x - 1) = -3(2y - 4)$ Use the distributive property.

$2x - 2 = -6y + 12$

$2x - 2 + 2 = -6y + 12 + 2$ Add 2 to both sides.

$2x = -6y + 14$

$2x + 6y = -6y + 6y + 14$ Add $6y$ to both sides to isolate 14.

Answer: $2x + 6y = 14$

Example 4

Write the following equation in standard form: $y = \dfrac{2}{5}x - 3$

$y = \dfrac{2}{5}x - 3$ Clear the fraction by multiplying by 5.

$5(y) = 5\left(\dfrac{2}{5}x\right) - 5(3)$ Remember to multiply **each term** by 5.

$5y = 2x - 15$ We need to move $2x$.

$-2x + 5y = -2x + 2x - 15$ Subtract $2x$ from both sides.

$-2x + 5y = -15$ Since the x-term is negative, multiply each term by -1.

$-1(-2x + 5y) = -1(-15)$ Distribute the -1.

Answer: $2x - 5y = 15$

Note: In standard form ($Ax + By = C$), we expect A to be a positive number.

Methods for Finding the Equation of a Line

Finding the Equation of a Line, Given Its Graph

The fastest way to find the equation of a line is to use what's called point-slope form.

Definition of **point-slope form:** The equation of a line that passes through the point (x_1, y_1) with slope m is $y - y_1 = m(x - x_1)$

Let's use this formula to find the equation of a line, given its graph.

Example 5

Find the equation of the line pictured below in slope-intercept form.

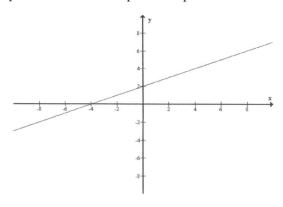

The line passes through the point $(-4, 0)$, which is the x-intercept, and the point $(0, 2)$, which is the y-intercept.

The slope of the line is $m = \dfrac{0 - 2}{-4 - 0} = \dfrac{-2}{-4} = \dfrac{1}{2}$ or 0.5. We can use either of the two points as our (x_1, y_1); we will pick $(0, 2)$, so $x_1 = 0$ and $y_1 = 2$.

$y - y_1 = m(x - x_1)$

$y - 2 = 0.5(x - 0)$ Substitute the given values and then use the distributive property.

$y - 2 = 0.5x$ Add 2 to both sides.

Answer: $y = 0.5x + 2$

Example 6

Find the equation of the line pictured below. Write your answer in standard form.

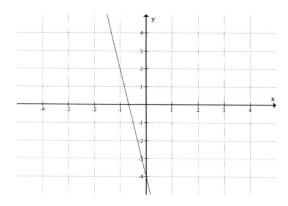

This line passes through the points $(0, -4)$ and $(-1, 2)$. (Pick the points on the graph that are the easiest to read.) The slope is $m = \dfrac{-4 - 2}{0 - (-1)} = -6$.

$$y - y_1 = m(x - x_1)$$

$y - (-4) = -6(x - 0)$ Substitute the given values, using $(0, -4)$ as the point, and then use the distributive property.

$y + 4 = -6x$

Remember that we want the answer in standard form; we need both x and y on the left side.

$y + 4 - 4 + 6x = -6x - 4 + 6x$ Subtract 4 from and add $6x$ to both sides.

Answer: $6x + y = -4$ Write in alphabetical order.

Example 7

Write the equation of the line pictured below in both forms.

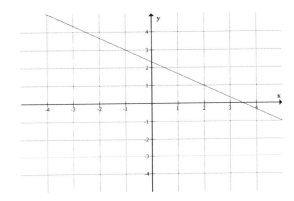

This line passes through the points $(2, 1)$ and $(-1, 3)$. The slope is $m = \dfrac{1 - 3}{2 - (-1)} = \dfrac{-2}{3}$

$$y - y_1 = m(x - x_1)$$

Use the point $(-1, 3)$ and the slope.

$y - 3 = \dfrac{-2}{3}(x - (-1))$

$y - 3 = \dfrac{-2}{3}x - \dfrac{2}{3}$ Use the distributive property.

$y - 3 + 3 = \dfrac{-2}{3}x - \dfrac{2}{3} + 3$ Add 3 to both sides. (Remember $3 = \dfrac{9}{3}$)

Answer: $y = \dfrac{-2}{3}x + \dfrac{7}{3}$

To write in standard form, do the following.

$3(y) = 3\left(\dfrac{-2}{3}x + \dfrac{7}{3}\right)$ Multiply by 3 to clear the fractions.

$3y = -2x + 7$ Add $2x$ to both sides.

Answer: $2x + 3y = 7$

Example 8

Find the equation of the line that passes through the point $(4, 2)$ with a slope of 3. Write your answer in slope-intercept form.

$y - y_1 = m(x - x_1)$

$y - (2) = 3(x - 4)$ Substitute the given values and then use the distributive property.

$y - 2 = 3x - 12$

$y - 2 + 2 = 3x - 12 + 2$ Add 2 to both sides.

Answer: $y = 3x - 10$

Example 9

Write the equation of the line that passes through the point $(-2, -5)$ with a slope of $\frac{1}{2}$. Write your answer in standard form.

$y - y_1 = m(x - x_1)$

$y - (-5) = \frac{1}{2}(x - (-2))$ Substitute the given values and then use the distributive property.

$y + 5 = \frac{1}{2}x + 1$ Subtract $-1/2x$ and 5 on both sides.

$-\frac{1}{2}x + y = -4$

$-2\left(\frac{-1}{2}x + y\right) = -2(-4)$ Multiply by -2 to clear the fractions.

Answer: $x - 2y = 8$

Note: Notice that we multiplied both sides by -2, not 2. If you multiply both sides by 2, you would obtain $-x + 2y = -8$. In standard form, however, we generally want the coefficient of x to be a positive number. That's why we multiplied both sides by -2.

FINDING THE EQUATION OF A LINE, GIVEN TWO POINTS

Just as we did above, we will use the point-slope form to find the equation of a line. The difference is that in the following examples we are told what the two points are. We do not have to find them on the graph; however, we do need to find the slope of the line between the given points in order to use the formula.

Example 10

Find the equation of the line that passes through the points $(2, 3)$ and $(-1, 4)$. Write your answer in both forms.

Find the slope: $m = \dfrac{(4)-(3)}{(-1)-(2)} = \dfrac{1}{-3} = -\dfrac{1}{3}$

$y - (3) = \dfrac{-1}{3}(x - (2))$ \qquad Substitute the point $(2, 3)$ and then use the distributive property.

$y - 3 = \dfrac{-1}{3}x + \dfrac{2}{3}$

$y - 3 + 3 = \dfrac{-1}{3}x + \dfrac{2}{3} + 3$ \qquad Add 3 to both sides.

Answer: $y = \dfrac{-1}{3}x + \dfrac{11}{3}$

To write in standard form:

$3(y) = 3\left(\dfrac{-1}{3}x + \dfrac{11}{3}\right)$ \qquad Multiply by 3 to clear fractions.

$3y = -1x + 11$ \qquad Add $1x$ to both sides.

Answer: $x + 3y = 11$

Example 11

Find the equations for the vertical and horizontal lines that pass through the point $(5, -3)$.

These are particularly easy. Recall that all vertical lines are of the form $x = c$ and all horizontal lines are of the form $y = c$, where c is a constant. The vertical line that passes through the point $(5, -3)$ is $x = 5$, which is the x-coordinate of that point. The horizontal line that passes through the point $(5, -3)$ is $y = -3$, which is the y-coordinate of that point.

Answer: The vertical line is $x = 5$.

The horizontal line is $y = -3$.

Example 12

Find the equation of the line that passes through the point $(2, 6)$ and is parallel to the line $3x + y = 4$. Write your answer in standard form.

We are looking for a second line that is parallel to the given line, $3x + y = 4$. Two lines are parallel if their slopes are equal, so we need to know the slope of $3x + y = 4$. Recall that the slope of $Ax + By = C$ is $\dfrac{-A}{B}$. In this case, $m = \dfrac{-3}{1} = -3$, so the slope of the line we are looking for is -3, and the line passes through the point $(2, 6)$.

$y - (6) = -3(x - (2))$ \qquad Substitute and use the distributive property.

$y - 6 = -3x + 6$

Chapter 3: Functions and Graphing

$y + 3x + 6 - 6 = -3x + 3x + 6 + 6$ Add 6 and $3x$ to both sides.

Answer: $3x + y = 12$ Write in alphabetical order.

Practice Problems

Write the following equations in slope-intercept form.

1. $3x + y = 5$
2. $-6x - y = 8$
3. $4x + 2y = 10$
4. $6x + 3y = -9$
5. $2x - 3y = 8$
6. $5x + 4y = 8$
7. $3(x - 1) = 2(y + 2)$
8. $4(y - 3) = -3(x - 5)$
9. $\frac{1}{2}x + 2y = 8$
10. $\frac{-1}{3}x + \frac{2}{3}y = -1$

Write the following equations in standard form.

11. $y = 4x - 2$
12. $y = -3x + 9$
13. $3x - 5y + 8 = -1$
14. $2x - 6y + 2 = 0$
15. $-2(y - 4) = -3(x - 1) + 4$
16. $6(2x + 3) = 4(y - 1) + 6$
17. $y = \frac{-2}{3}x + 4$
18. $y = \frac{3}{4}x + \frac{1}{2}$

Find the equation of the line pictured in the following graphs. Write your answer in slope-intercept form.

19.

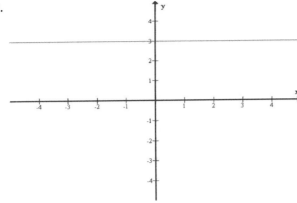

20.

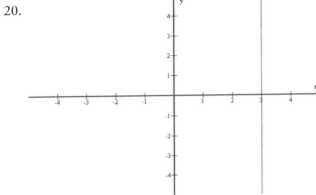

21.

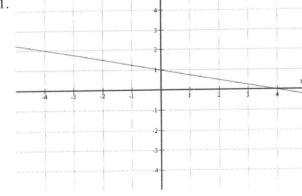

Chapter 3: Functions and Graphing

22.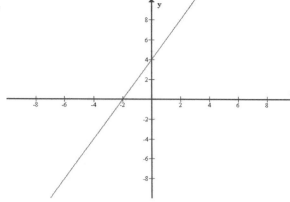

23.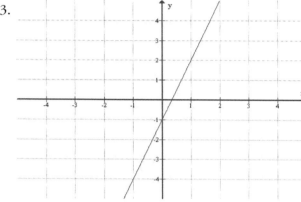

24.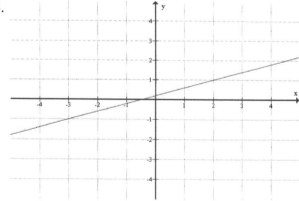

Find the equation of the line, given the following information. Write your answer in slope-intercept form.

25. The line passes through the point $(-3, 4)$ with a slope of -5.

26. The line passes through the point $(4, 1)$ with a slope of 3.

27. The line passes through the point $(-1, -2)$ with a slope of $\dfrac{2}{3}$.

28. The line passes through the point $(2, -3)$ with a slope of $\dfrac{-1}{3}$.

29. The line passes through the point $(3, 7)$ with a slope of 0.
30. The line passes through the point $(0, 1)$ with no slope.

Find the equation of the line, given the following information. Write your answer in standard form.

31. The line passes through the points $(0, 2)$ and $(4, 5)$.
32. The line passes through the points $(-1, 3)$ and $(4, -2)$.
33. The line passes through the points $(0, -3)$ and $(2, 6)$.
34. The line passes through the points $(-4, 2)$ and $(3, -1)$.

Find the equations of the lines, given the following information. Write your answers in slope-intercept form.

35. Find the equations of the vertical and horizontal lines that pass through the point $(0, 2)$.
36. Find the equations of the vertical and horizontal lines that pass through the point $(-3, 1)$.
37. Find the equation of the line that passes through $(5, 1)$ and is parallel to the line $2x + y = 6$.
38. Find the equation of the line that passes through $(-3, 4)$ and is parallel to the line $5x - y = 10$.
39. Find the equation of the line that passes through $(3, -1)$ and is parallel to the line $3x - 5y = 7$.
40. Find the equation of the line that passes through $(-1, -2)$ and is parallel to the line $5x - 2y = 4$.
41. Find the equation of the line that passes through $(4, 1)$ and is perpendicular to the line $2x + y = 9$.
42. Find the equation of the line that passes through $(3, 4)$ and is perpendicular to the line $5x - y = 6$.
43. Find the equation of the line that passes through $(-1, 0)$ and is perpendicular to the line $2x + 3y = 5$.
44. Find the equation of the line that passes through $(6, -2)$ and is perpendicular to the line $4x - 3y = 6$.

Section 4
Graphing Inequalities in Two Dimensions

Learning Objectives

When you finish your study of this section, you should be able to
- Graph a linear inequality in the coordinate plane

Inequalities in Two Dimensions

In this section, we will look at graphing inequalities in the two-dimensional coordinate plane. You have probably at some time graphed an inequality in one dimension. For example, a one-dimensional graph of the inequality $x < 2$ follows:

In this section, we will look at graphing these types of inequalities in two dimensions. Before we start, we should review the equations of horizontal and vertical lines:
A horizontal line is written as $y = c$, where c is any real number.
A **vertical line** is written as $x = c$, where c is any real number.
Let's look at some examples.

Example 1

Graph $y \leq -3$ in two dimensions.

First, $y = -3$ is a horizontal line passing through the point $(0, -3)$ on the y-axis.

Since $y \leq -3$ includes the possibility that $y = -3$, we will draw the line as a **solid** line.

Since we are interested in values of y that are less than or equal to -3, we shade **below** the line. The graph of $y \leq -3$ follows:

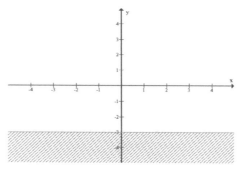

Chapter 3: Functions and Graphing

Example 2

Graph $y > 5x - 2$.

First, we graph $y = 5x - 2$. To do this, we find the y-intercept.

If $x = 0$, $y = 5(0) - 2 = -2$. The y-intercept is $(0, -2)$.

Next, we find the x-intercept:

If $y = 0$, $0 = 5x - 2$	Add 2 to both sides.
$2 = 5x$	Divide both sides by 5.
$0.4 = x$	The x-intercept is $(0.4, 0)$

$y > 5x - 2$ does **not** include the possibility that $y = 5x - 2$, so we draw the line as a **dotted** line. Next, we just figure out which side of the line to shade. To do this, we recommend testing the point $(0,0)$. That is, we substitute the value 0 for both x and y in the inequality we have been given. If we end up with a **true** sentence, we shade the side of the line containing $(0,0)$. If we end up with a **false** sentence, we shade the side of the line that does not contain $(0,0)$.

In this problem, if $x = 0$ and $y = 0$, we are saying that $0 > 5(0) - 2$; that is, $0 > -2$, a sentence which is certainly true. Thus, we shade the side of the line containing $(0,0)$. To make sure we haven't made a mistake, let's test a point that is not in the shaded region to determine if its substitution makes the inequality a false sentence. We will test the point $(3,0)$:

If $x = 3$ and $y = 0$, we are saying that $0 > 5(3) - 2$; that is, $0 > 13$, a sentence which is certainly false. The graph of $y > 5x - 2$ follows. We have put dots at the points $(0,0)$ and $(3,0)$ to make them easier to see.

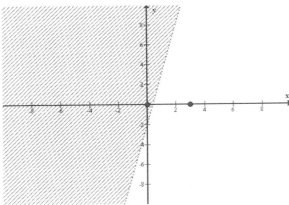

Example 3

Graph $2x + 3y \leq 10$.

To graph the line, we find the x-intercept and y-intercept. If $y = 0$, $2x = 10$, so $x = 5$. The x-intercept is $(5,0)$. If $x = 0$, $3y = 10$, so $y = 10/3$ or roughly 3.3. The y-intercept is $(0, 3.3)$.

The inequality contains an equal sign, so we draw a solid line.

Finally, we test $(0,0)$.

If $x = 0$ and $y = 0$, we are saying that $2(0) + 3(0) \leq 10$; that is, $0 \leq 10$, a sentence which is certainly true. Thus, we shade the side of the line that contains $(0,0)$.

The graph of $2x + 3y \leq 10$ follows:

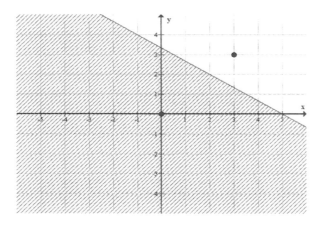

You should verify for yourself that the point $(3, 3)$, shown on the graph, makes the inequality false.

Example 4

Graph $y \geq \dfrac{-1}{3}x - 2$.

To graph the line, substitute values for x.

(Hint: Use multiples of the denominator of the fraction so that the arithmetic will work out easily.)

If $x = 0$, $y = \dfrac{-1}{3}(0) - 2 = -2$. Thus, one point on the line is $(0, -2)$.

Now, let $x = 3$ so that $y = \dfrac{-1}{3}(3) - 2 = -1 - 2 = -3$. Thus, another point on the line is $(3, -3)$.

The inequality includes an equal sign, so we draw a solid line. Finally, checking the point $(0, 0)$, we find that $0 \geq \dfrac{-1}{3}(0) - 2$, which means that $0 \geq -2$, a sentence which is certainly true.

Thus, we shade the side of the line that contains $(0, 0)$. The graph of $y \geq \dfrac{-1}{3}x - 2$ follows:

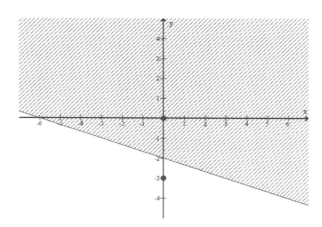

You should verify for yourself that the point $(0, -3)$, shown on the graph, makes the inequality false.

Example 5

Graph $x + y < 0$.

In this problem, both the x-intercept and the y-intercept are the point $(0,0)$. (Verify this.) To draw the line, we need at least one more point, so let $x = 1$. Then, $1 + y = 0$, a sentence which implies that $y = -1$, so a second point on the line is $(1, -1)$. The inequality does not include an equal sign, so we draw the line as a dotted line. Since the line passes through the point $(0,0)$, we cannot use it as our test point. We must use a different point, for example, $(0,2)$. If $x = 0$ and $y = 2$, then we are saying that $0 + 2 < 0$, that is, that $2 < 0$, a sentence which is certainly false. Thus, we shade the side of the line that does not contain the point $(0,2)$. The graph of $x + y < 0$ follows:

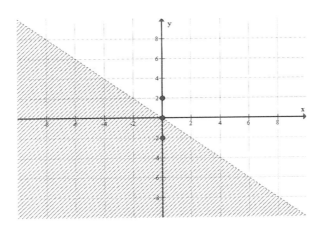

You should verify for yourself that $(0, -2)$, the third point shown on the graph, makes the inequality a true sentence.

Practice Problems

Graph each of the following on a coordinate plane.

1. $x \leq 2$
2. $x > -1$
3. $y < 5$
4. $y \leq 3$
5. $y > 4x - 1$
6. $y < -3x + 5$
7. $y \leq x + 2$
8. $y \geq -x - 4$
9. $y \leq \frac{1}{2}x$

10. $y > \dfrac{-3}{4}x + 2$

11. $2x - 3y < 6$

12. $4x + y > -2$

13. $5x - 10y \leq 15$

14. $3x + 3y \geq 9$

15. $x - y > 0$

16. $2x + y \leq 0$

SECTION 5
MORE ON FUNCTIONS

Learning Objectives

When you finish your study of this section, you should be able to
- Determine if a set of ordered pairs is a function
- Understand and use function notation
- Determine if a given graph represents a function
- Determine the domain and range of a relation or function
- Graph various functions

Relations and Functions

Any set of ordered pairs (points) is a **relation**. A **function** is a set of ordered pairs no two of which contain the same first component. In other words, each x-value is only allowed to go to a single y-value. Let's look at some examples.

Example 1

Does $\{(3,2), (4,5), (6,5), (7,6)\}$ represent a function?

Look at the four points. Do you see an x-coordinate that appears twice? No. Thus, the set is a function. We are also interested in the **domain** of the function, a list of all of its x-coordinates, and the **range** of the function, a list of all of its y-coordinates.

The domain is $\{3, 4, 6, 7\}$.

The range is $\{2, 5, 6\}$.

Example 2

Is $\{(-2,0), (0,4), (0,6), (2,1)\}$ a function?

Look at the four points. Do you see an x-coordinate that appears twice? Yes. We have the points $(0,4)$ and $(0,6)$. If 0 "goes to" 4, it is not allowed to also "go to" 6. Thus, the set is not a function; however, it is a relation.

Its domain is $\{-2, 0, 2\}$.

Its range is $\{0, 1, 4, 6\}$.

We use lower-case letters to represent functions; the letters f, g, and h are popular. Thus, $f(x)$ means that we are calling our function f and that the variable in this function is x. The choice of letter really makes no difference. Thus, $f(x) = x^2 + 5x - 1$ is really the same function as $f(t) = t^2 + 5t - 1$.

In a function, x is called the **independent variable** and $f(x)$ is called the **dependent variable** because it depends on the value of x.

To change an equation into function notation, simply solve for y and replace y with $f(x)$.

Example 3

Change $10x + 5y = 25$ into function notation.

If $10x + 5y = 25$, then $5y = 25 - 10x$, so $y = 5 - 2x$.

Answer: $f(x) = 5 - 2x$

Example 4

If $f(x) = 3x^2 - 2x$, find $f(2)$, $f(-1)$, and $f(r^2)$.

Function notation essentially tells us what number (or expression) to substitute for x.

Answer: $f(2) = 3(2)^2 - 2(2) = 12 - 4 = 8$.

$f(-1) = 3(-1)^2 - 2(-1) = 3 + 2 = 5$

$f(r^2) = 3(r^2)^2 - 2(r^2) = 3r^4 - 2r^2$

Vertical Line Test

If you look at the graph of a relation, you can decide if it is a function by using the **vertical line test**. If at any place on the graph, you can draw a vertical line, and that vertical line intersects the graph more than once, the graph does **not** represent a function. To relate this to the definition of a function, if a vertical line passes through a graph twice, then at that point you have shown that the same x value has two different y values, which means it is not a function.

Example 5

Does the following graph represent a function? Also, find its domain and range.

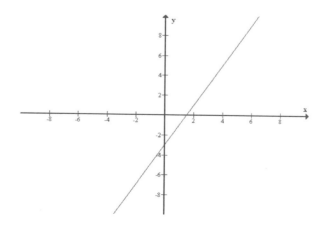

Chapter 3: Functions and Graphing

This graph passes the vertical line test, so it is a function.

The graph consists of a straight line that continues forever both upwards and downwards to the left and right, the domain is all real numbers, and the range is all real numbers. Using interval notation, all real numbers is represented as $(-\infty, \infty)$.

Note: When we are trying to find the domain, we look to see how far the graph extends to the left and right. When we are trying to find the range, we look to see how far the graph extends upwards and downwards.

Example 6

Does the graph below represent a function? Find its domain and range.

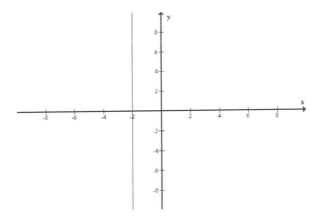

This graph fails the vertical line test (it **is** a vertical line), so the graph does not represent a function, though it is a relation. For every point on this line, the x-coordinate is -2, so, the domain is $\{-2\}$. However, the vertical line does continue forever both upwards and downwards, so, the range is all real numbers, or $(-\infty, \infty)$.

Example 7

Does the following graph represent a function? Find its domain and range.

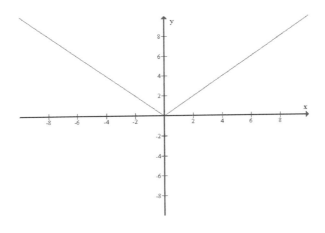

This graph passes the vertical line test, so it is a function. This graph extends to the left and right forever, so its domain is all real numbers, or $(-\infty, \infty)$. However, the lowest point on this graph is the point $(0,0)$, so the range is $[0, \infty)$.

Example 8

Does the graph below represent a function? Find its domain and range.

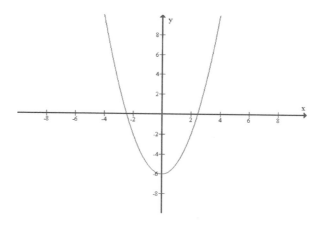

This graph passes the vertical line test, so it is a function. This graph extends to the left and right forever, so its domain is all real numbers, or $(-\infty, \infty)$. However, the lowest point on the graph is the point $(0, -6)$, so the range is $[-6, \infty)$. (Note that we look at the y-coordinate of the point when finding the range, not the x-coordinate.)

GRAPHING FUNCTIONS

To draw the graph of a function, we will create a simple chart like we did when we were graphing lines. Once we have plotted a few points, we draw a smooth curve through the points to represent the function. The functions we will study in this section include the following:

Type of Function	General Form
Linear	$f(x) = mx + b \, (m \neq 0)$
Constant	$f(x) = c$
Square Root	$f(x) = \sqrt{x}$
Absolute Value	$f(x) = |x|$
Quadratic	$f(x) = ax^2 + bx + c$

Example 9

Graph $f(x) = -3x + 1$.

x	$f(x)$	Point
0	$-3(0) + 1 = 1$	$(0, 1)$
1	$-3(1) + 1 = -2$	$(1, -2)$

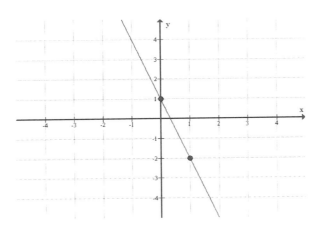

Example 10

Graph $f(x) = \sqrt{x - 2}$.

x	$f(x)$	Point
2	$\sqrt{2-2} = 0$	$(2, 0)$
3	$\sqrt{3-2} = 1$	$(3, 1)$
6	$\sqrt{6-2} = 2$	$(6, 2)$

The graph of $f(x) = \sqrt{x-2}$ follows:

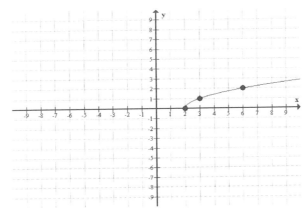

You may wonder why we started with 2 in the chart above. Recall that the smallest number we can take the square root of is 0. Thus, we figure out what value of x would make $x - 2 = 0$. That value of x is 2. We picked 3 and 6 so that the square roots would be whole numbers. If you are using a calculator, you could have used any numbers larger than 2, and your calculator would have given you an approximate square root.

Example 11

Graph $f(x) = |x + 2|$.

When graphing an absolute value function, start with the x-value that makes the expression inside the absolute value sign equal to 0. In this case, that x-value is -2. Then, pick two numbers on either side of that x-value. Five points will be sufficient to graph the v-shaped function.

x	$f(x)$	Point
0	$|0 + 2| = 2$	$(0, 2)$
-1	$|-1 + 2| = 1$	$(-1, 1)$
-2	$|-2 + 2| = 0$	$(-2, 0)$
-3	$|-3 + 2| = 1$	$(-3, 1)$
-4	$|-4 + 2| = 2$	$(-4, 2)$

The graph of $f(x) = |x + 2|$ follows:

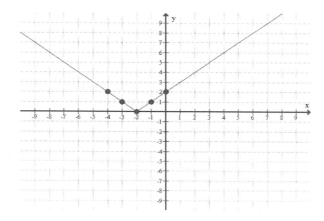

Example 12

Graph $f(x) = x^2 - 1$.

Note: When graphing a second degree function, that is, a function in which the largest exponent is 2, if you pick $x = -2, -1, 0, 1$, and 2 as x-values, you will always have plenty of points to draw an accurate sketch of the function. Second degree functions have this U-shaped graph, sometimes opening up, sometimes opening down. The graph of such functions is called a **parabola**.

x	$f(x)$	Point
-2	$(-2)^2 - 1 = 3$	$(-2, 3)$
-1	$(-1)^2 - 1 = 0$	$(-1, 0)$
0	$0^2 - 1 = -1$	$(0, -1)$
1	$1^2 - 1 = 0$	$(1, 0)$
2	$2^2 - 1 = 3$	$(2, 3)$

The graph of $f(x) = x^2 - 1$ follows:

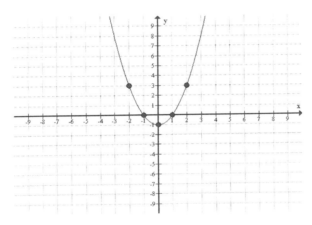

Example 13

Graph $f(x) = -x^2 - 1$.

x	$f(x)$	Point
-2	$-(-2)^2 - 1 = -5$	$(-2, -5)$
-1	$-(-1)^2 - 1 = -2$	$(-1, -2)$
0	$-(0)^2 - 1 = -1$	$(0, -1)$
1	$-(1)^2 - 1 = -2$	$(1, -2)$
2	$-(2)^2 - 1 = -5$	$(2, -5)$

Note: Notice that after each x-value is squared, making it a positive, the negative on the outside attaches to it, making it a negative.

The graph of $f(x) = -x^2 - 1$ follows:

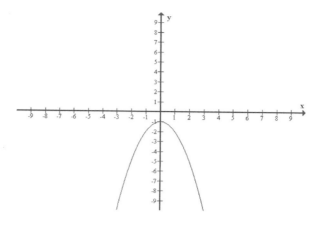

Example 14

Graph $f(x) = -|x + 2|$.

x	$f(x)$	Point
0	$-|0 + 2| = -2$	$(0, -2)$
-1	$-|-1 + 2| = -1$	$(-1, -1)$
-2	$-|-2 + 2| = -0$	$(-2, 0)$
-3	$-|-3 + 2| = -1$	$(-3, -1)$
-4	$-|-4 + 2| = -2$	$(-4, -2)$

Note: This graph is just like the function in Example 11, with the exception of the negative in front. Notice that the negative flips the original graph upside down.

The graph of $f(x) = -|x + 2|$ follows:

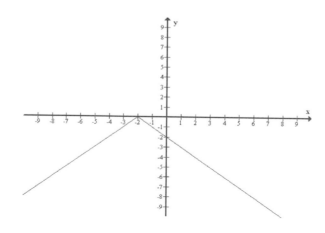

Practice Problems

Determine whether each relation is a function. In either case, find the domain and range.

1. $(3,4), (4,5), (5,6), (6,7)$
2. $(-1,0), (-1,3), (2,4), (2,6)$
3. $(4,5), (5,5), (6,5)$
4. $(.3,0), (.4,0), (.4,1), (.5,3)$

Change the following equations into function notation.

5. $y = 4x - 3$
6. $3x + y = 8$
7. $3x - 3y = 9$
8. $8x + 2y = 4$

Find the values $f(2)$, $f(-3)$, and $f(q^2)$ for the following functions:

9. $f(x) = 4x - 5$
10. $f(x) = -2x + 8$
11. $f(x) = 3x^2 - 5x$
12. $f(x) = 2x^2 - 6x + 7$
13. $f(x) = \sqrt{x - 1}$
14. $f(x) = |x - 4|$

Determine which of the following graphs represent functions.

15.

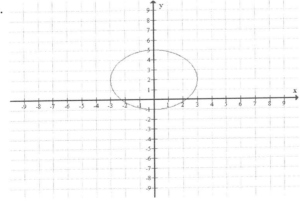

Example 2

Solve: $x + 2y = -1$; $x - y = 2$

First, look at $x + 2y = -1$:

To find the x-intercept: If $y = 0$, $x + 2(0) = -1$, so $x = -1$. The x-intercept is $(-1, 0)$

To find the y-intercept: If $x = 0$, $0 + 2y = -1$, so $y = -0.5$. The y-intercept is $(0, -0.5)$

(*Even though this number is a decimal, it is the only solution you can graph precisely, so you can use it.*)

Next, look at $x - y = 2$:

To find the x-intercept: If $y = 0$, $x - 0 = 2$, so $x = 2$.

The x-intercept is $(2, 0)$

To find the y-intercept: If $x = 0$, $0 - y = 2$, so $y = -2$.

The y-intercept is $(0, -2)$

Sketch the two lines on the same graph:

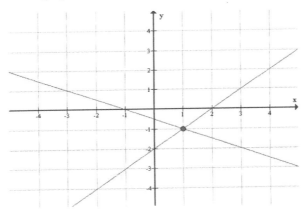

The lines intersect at the point $(1, -1)$, so the solution to the system of equations is $x = 1$ and $y = -1$.

Answer: $x = 1$ and $y = -1$

Example 3

Solve: $y = 4x - 1$; $y = 4x + 3$

Before you start graphing, notice that these two lines have the same slope: 4. What do you know about two lines that have the same slope? That's right, they are parallel. Two lines that are parallel will never intersect. Thus, this system of equations has no solution. In case you are not convinced, see the graph of the two lines on the following graphic:

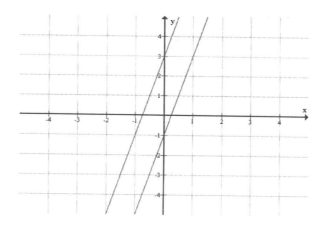

Answer: no solution

SOLVING SYSTEMS OF EQUATIONS BY USING THE SUBSTITUTION METHOD

A second way to solve systems of equations is the substitution method. In this method, we solve one equation for either x or y and then substitute that expression into the other equation. An example will make this procedure clearer.

Example 4

Solve: $x + y = 7$; $y = 3x - 5$

Notice that the second equation has already been solved for y. We know that $y = 3x - 5$. We substitute that expression for y in the first equation:

$x + (3x - 5) = 7$	Substitute $3x - 5$ for y. in the other equation
$4x - 5 = 7$	Combine like terms.
$4x = 12$	Divide both sides by 4.
$x = 3$	

To find the value of y, we let $x = 3$ in the first equation:

$3 + y = 7$	Subtract 3 from both sides.
$y = 4$	

Answer: $x = 3$ and $y = 4$

Note: We could have substituted $x = 3$ into the second equation to find out that $y = 4$. Either equation will work.

Example 5

Solve: $x - 4y = 15$; $3x + 3y = -15$

In this problem, neither equation is already solved, so we must pick one to solve. The first equation looks a little less complicated than the second equation because the x-value in the first equation has a coefficient of 1, so we will solve the first equation for x:

$x - 4y + 4y = 15 + 4y$	Add $4y$ to both sides.
$x = 15 + 4y$	

Now, we substitute that expression for x in the second equation:

$3(15 + 4y) + 3y = -15$	Substitute.
$45 + 12y + 3y = -15$	Use the distributive property.
$45 + 15y = -15$	Combine like terms.
$15y = -60$	Divide both sides by 15.
$y = -4$	

Finally, since we know that $y = -4$, we substitute -4 for y in the first equation:

$x - 4(-4) = 15$	Substitute.
$x + 16 = 15$	Subtract 16 from both sides.
$x = -1$	

Note: We could also have substituted -4 for y in the second equation to solve for x, but the arithmetic looked a bit easier to do in the first equation.

Answer: $x = -1$ and $y = -4$

SOLVING SYSTEMS OF EQUATIONS BY USING THE ELIMINATION METHOD

Since some equations intersect at points that are not whole numbers, and because some equations are too messy to use the substitution method, we have what we call the "fool-proof method" that works for any system of two equations: the elimination method. This method involves the addition of the two equations in order to cancel out one of the two variables, thus allowing you to solve for the remaining variable. Before you add the two equations together, it is normally necessary to alter one or both of the equations so that, when added, one of the variables cancels out. To use this method, follow these steps:

Step 1: Look at the two equations and determine if, when added, one of the two variables will cancel out to zero.

Step 2: If the two equations will not cancel to zero when added together, then multiply one or both of the equations by a number (not zero) such that the coefficients of either the x or y variable have opposite signs (for example, 5 and -5).

Step 3: Add the two equations together.

Step 4: Solve the remaining equation for the variable that did not cancel out.

Step 5: Substitute the value for x or y obtained in step 4 in either equation to find the other variable.

Step 6: Check to see that the ordered pair satisfies both equations.

It's not as hard as it sounds. Let's try an example.

Example 6

Solve: $x - 4y = 15$; $3x + 3y = -15$

You may recognize this problem as Example 5. Using the elimination method, we are going to redo the problem so that you see we still obtain the same final answer. First, we write the two equations vertically, one above the other:

$x - 4y = 15$

$3x + 3y = -15$

Step 1: Neither variable will cancel out if we add the two equations together; thus, we move to step 2.

Step 2: $3x$ in the second equation is a multiple of x in the first equation, so we multiply **both sides** of the first equation by -3, which is the opposite of 3.

$3(x - 4y = 15)$ Distribute the -3 on **both** sides of the equation.

$-3x + 12y = -45$ This is the new first equation.

Since we multiplied both sides of an equation by a constant, the two equations are equivalent; that is, any values of x and y that make the first equation a true sentence will also make the second equation a true sentence.

Step 3: We add the new first equation to the original second equation:

$-3x + 12y = -45$

$3x + 3y = -15$

$15y = -60$

Step 4: We solve the remaining equation for y. By dividing both sides by 15, we obtain $y = -4$.

Step 5: We substitute -4 for y in either equation. We will use the first:

$x - 4(-4) = 15$ Substitute.

$x + 16 = 15$ Subtract 16 from both sides.

$x = -1$

Answer: $x = -1$ and $y = -4$. (Notice that this is the same answer we obtained in Example 5, so we will not check the answer.)

Example 7

Solve: $4x + 2y = -12$; $6x + 3y = -12$

Step 1: Write the two equations vertically, one above the other, to determine if you can add them together to cancel out a variable:

$4x + 2y = -12$

$\underline{6x + 3y = -12}$

Step 2: Since neither variable will cancel out through addition, you need to find a number, or numbers, to multiply by the equations to cancel out the x or y. There does not appear to be a number to multiply by just one equation to cancel out the x or y, so you need to find a number that both 4 and 6 divide into evenly to cancel the x-values out of the system of equations. A simple choice would be 12. Multiply the top equation by -3 and the bottom equation by 2 so that the x-terms will add to 0.

$-3(4x + 2y = -12)$	Distribute the -3 on both sides of the first equation.
$2(6x + 3y = -12)$	Distribute the 2 on both sides of the second equation.
$-12x - 6y = 36$	This is the new first equation.
$12x + 6y = -24$	This is the new second equation.

Step 3: Add the equations.

$-12x - 6y = 36$

$\underline{12x + 6y = -24}$

$ 0 = 12$

Step 4: Notice that something very strange happened. Both variables added to 0. The final sentence, $0 = 12$, is false. This means that this system of equations has no solution. Simultaneous equations that have no common solution are called inconsistent equations.

Answer: no solution or inconsistent

If you check, you should be able to verify that both lines have the same slope, -2. This means that they are parallel lines, which never intersect, and that the two equations are inconsistent.

Example 8

Solve: $3x - 2y = -7$; $4x + 4y = -6$

Step 1: Write the two equations, one above the other, and determine if you can cancel a variable through addition.

$3x - 2y = -7$

$\underline{4x + 4y = -6}$

Step 2: Since you can't cancel a variable through addition, look to see if one of the variables is a multiple of the other. In this case, $4y$ is a multiple of $2y$, so eliminate y this time. If you add $-4y$ to the $4y$ in the second equation, you would obtain 0. To do that, you need to multiply the first equation by 2:

$2(3x - 2y = -7)$ Multiply **both** sides by 2 to obtain the

$6x - 4y = -14$ new first equation.

Step 3: Add the new first equation to the original second equation.

$6x - 4y = -14$

$\underline{4x + 4y = -6}$

$10x = -20$

Step 4: Solve the new equation by dividing both sides by 10; the solution is $x = -2$.

Step 5: Substitute -2 for x in the first equation:

$3(-2) - 2y = -7$ Substitute.

$-6 - 2y = -7$ Add 6 to both sides.

$-2y = -1$

$y = \dfrac{-1}{-2} = \dfrac{1}{2}$ Divide both sides by -2.

Answer: $x = -2$ and $y = \dfrac{1}{2}$

Example 9

Solve: $3x - y = 6$; $\dfrac{1}{2}x + \dfrac{1}{3}y = 1$

Yuck! Fractions!

Don't panic. You first need to change the fractional equation into standard form. Call this a preliminary step before Step 1.

Multiply both sides of the second equation by $6 (2 \cdot 3 = 6)$ to clear the fractions.

$$6\left(\frac{1}{2}x + \frac{1}{3}y = 1\right)$$

$3x + 2y = 6$

Now, solve the system of equations:

$3x - y = 6$

$3x + 2y = 6$

Step 1: If you add the equations together, a term will not cancel out; however, they are very close to being opposites.

Step 2: Change the first term, $3x$, to $-3x$ by multiplying both sides by -1.

$-1(3x - y = 6)$ Multiply.

$-3x + y = -6$ This is the new first equation.

Step 3: Add the equations together.

$-3x + y = -6$
$\underline{3x + 2y = 6}$
$3y = 0$

Step 4: Solve the remaining equation for y by dividing both sides by 3; the solution is $y = 0$.

Step 5: Since $y = 0$, substitute 0 for y in the first equation.

$3x - (0) = 6$ Substitute.

$3x = 6$ Divide both sides by 3.

$x = 2$

Answer: $x = 2$ and $y = 0$

Example 10

Solve: $x + 3y = 5$; $3x + 9y = 15$

Step 1: Rewrite the equations, one above the other.

$x + 3y = 5$

$3x + 9y = 15$

Step 2: Since you can't cancel a variable through addition, look to see if one of the variables is a multiple of the other. In this case, $3x$ is a multiple of x, so eliminate the x this time. (You may have also noticed that $9y$ is a multiple of $3y$, which makes this problem unique).

$-3(x + 3y) = 5$ Multiply both sides of the equation by -3.

$-3x - 9y = -15$ New first equation

Step 3: Add the new first equation to the second equation.

$-3x - 9y = -15$
$\underline{3x + 9y = 15}$ Notice everything on both sides cancels out to zero.
$0 = 0$

Step 4: Notice that something strange happened. Just like Example 7, both variables summed to 0; however, in this problem, the right side added to zero also. Since the statement $0 = 0$ is true, the answer is that there are infinitely many solutions. In other words, these are actually the same equation (try graphing them and see!) and any solution that works for one also works in the other.

Answer: infinite many solutions

Practice Problems

Solve the following systems of equations graphically, if possible

1. $x + y = 2; -2x + y = 2$
2. $2x + 2y = -4; -3x - 6y = 9$
3. $y = x + 3; y = 2x + 1$
4. $y = -4x + 5; y = x - 5$

Solve the following systems of equations by using the substitution method.

5. $x + 3y = 0; 2x - y = -7$
6. $3x + y = -8; 2x - 2y = -8$
7. $4x + 3y = 14; -2x + y = 5$
8. $-2x + y = 4; 2x - 2y = 4$
9. $3x + 4y = 4; y = x + 1$
10. $2x + 4y = 8; x + 2y = 7$

Solve the following systems of equations by using the elimination method.

11. $2x + 3y = 10; 4x + y = 10$
12. $3x - 2y = -15; 2x + 6y = -10$
13. $5x - 3y = -1; 3x + 6y = 2$
14. $x + 2y = 12; 2x - 3y = 10$
15. $x + 2y = 5; 2x + 4y = 10$
16. $3x - 4y = -14; 2x + 3y = -15$

17. $2x + 3y = -14$; $4x - 2y = 12$

18. $3x - y = 8$; $6x - 2y = 10$

19. $\frac{1}{3}x + \frac{2}{3}y = 2$; $x + 2y = 6$

20. $\frac{1}{2}x + \frac{1}{4}y = -2$; $\frac{3}{2}x - \frac{3}{4}y = 0$

Section 2
Word Problems Involving Systems of Equations

Learning Objectives

When you finish your study of this section, you should be able to solve word problems involving
- Sums and differences of numbers
- Investments
- Purchases
- Mixtures

Most of you probably don't like word problems, so this may not be a section you are looking forward to. However, we think you will find that you can do the word problems that we are about to discuss. Why do we think that? Because word problems are a bit easier to set up when you use two variables instead of one. Let's see if we can convince you.

Solving Word Problems Involving Sums and Differences

For word problems involving sums and differences, the goal is to write out two equations from the words given (Chapter 2, Section 3). These equations will each contain two unknowns, which will allow you to solve the equations by the same means used in the first section of this chapter. You will notice that we always choose the "fool-proof" elimination method to solve these problems. Let's give one a try.

Example 1

The sum of two numbers is -3. Their difference is 27. Find the two numbers.

We are looking for two numbers; let's call them x and y. Their sum is -3, so $x + y = -3$. Their difference is 27, so $x - y = 27$. Let's solve that system of equations.

$x + y = -3$ We will use the elimination method. Add.
$x - y = 27$

$2x = 24$ Divide both sides by 2.

$x = 12$

Since $x = 12$, we will substitute this value into the first equation to solve for y.

$12 + y = -3$ Subtract 12 from both sides.

$y = -15$

Answer: The numbers are 12 and -15.

Always check your answer to a word problem.

$12 + (-15) = -3$ Their sum adds to -3.

$12 - (-15) = 27$ Their difference is 27.

Solving Word Problems Involving Investments

These problems involve investing money in accounts using simple interest. Recall that the simple interest formula is interest = (principal)(rate)(time) or $I = Prt$. Let's try a problem to see how the formula is used in the context of this book.

Example 2

Bill invested $30,000 and earned $1,050 in interest after one year. If some of the money earned 3% and the remainder 6%, then how much did he invest in each account?

(My co-author seems to think I have $30,000 lying around!)

In these problems, we have a large amount of money to be broken up into two accounts. Let's call the amount in the 3% account x and the amount in the 6% account y. Since the total amount of money is $30,000, we know that $x + y = 30000$. This will be our first equation.

To find a second equation, we need to address how simple interest is calculated. We know one account is earning 3%, and Bill has invested x dollars in this account. To find the sum of money earned in this account, we would substitute the following in the interest formula:

$I = x(.03)(1)$ or $I = .03x$.

For the 6% account, we would substitute the following.

$I = y(.06)(1)$ or $I = .06y$.

Since the total amount of interest is $1,050, we know that $.03x + .06y = 1050$. This is the second equation we need in order to solve the system. Now, let's solve it.

$x + y = 30000$
$.03x + .06y = 1050$

Let's multiply the first equation by $-.03$, so that the x-terms will cancel out.

$-.03(x + y) = -.03(30000)$

$-.03x - .03y = -900$ We obtained our new first equation.

$-.03x - .03y = -900$ Add the two equations together.
$\underline{.03x + .06y = 1050}$

$.03y = 150$	Divide both sides by .03.
$y = 5000$	This represents the amount of money in the 6% account.

Since $y = 5000$, we will substitute 5000 into the first equation to solve for x:

$x + 5000 = 30000$	Subtract 5000 from both sides.
$x = 25000$	

Answer: Bill invested $25,000 at 3% and $5,000 at 6%.

Check the answer:

$25000(.03) + 5000(.06) = 1050$	The interest earned checks out.
$25000 + 5000 = 30000$	The total investment amount checks out.

Solving Word Problems Involving Purchases

These problems are quite similar to the previous ones. In other words, we want to create a system of two equations with two unknowns and then solve the system. Let's go right to an example.

Example 3

Charlie went to his local fishing products store and bought four lures and five traps for $53.00. The next time he went to the store he bought two lures and three traps for $29.00. How much do lures and traps cost? (Assume that he was purchasing these items at the same price both times.)

Let x represent the price of a lure and y represent the price of a trap. For his first trip, we are given that he bought four lures, so the cost of four lures can be written as 4 • (the cost of one lure) or $4x$. We were also told that he bought five traps. This can be written as 5 • (the cost of one trap) or $5y$.

Next, to write the complete equation for the first purchase, we would add the total spent for the lures to the total spent for the traps:

(total spent for the lures) + (total spent for the traps) = total spent on the trip

$$4x + 5y = 53$$

Thus, the first equation is $4x + 5y = 53$.

Now, we will set up the second equation the same way.

(total spent for the lures) + (total spent for the traps) = total spent on the trip

$$2x + 3y = 29$$

Thus, the second equation is $2x + 3y = 29$.

Let's solve the system.

$4x + 5y = 53$

$2x + 3y = 29$

Multiply the second equation by -2.

$-2(2x + 3y) = -2(29)$

$-4x - 6y = -58$ We obtained our new second equation.

$-4x - 6y = -58$ Add the two equations together.
$4x + 5y = 53$

$-y = -5$ Divide both sides by -1.

$y = 5$ This represents the cost of the traps.

Since $y = 5$, we can use the first equation to solve for x.

$4x + 5(5) = 53$

$4x + 25 = 53$ Subtract 25 from both sides.

$4x = 28$ Divide both sides by 4.

$x = 7$

Answer: lures cost $7, traps cost $5

Check the answer.

$4(7) + 5(5) = 53$ The first purchase checks out.

$2(7) + 3(5) = 29$ The second purchase checks out also.

Solving Word Problems Involving Mixtures

These problems are also similar to the investment problems in that percents are involved. Remember that if you want to take a percent of a number, you multiply by the decimal equivalent of the percent. In these problems, our goal again is to set up two equations with two unknowns (x and y), and then solve. Let's go right to an example.

Example 4

A chemist wants to mix a 2% acidic solution with a 10% acidic solution in order to obtain 20 liters of an 8% acidic solution. How much of each solution will be needed?

Let x be the number of liters of 2% solution and y be the number of liters of 10% solution. We know that $x + y = 20$ liters total. This is our first equation:

$x + y = 20$

The formula for the second equation looks like this:

(number of liters)(percentage) + (number of liters)(percentage) = (total liters)(percentage)

Remember that to find a percent of a number, you multiply the percent by the number. This gives us our second equation:

$.02x + .10y = 20(.08)$ or

$.02x + .10y = 1.6$

Now, let's solve the system.

$x + y = 20$

$.02x + .10y = 1.6$

Let's multiply the first equation by $-.02$, so that the x-terms will cancel:

$-.02(x + y) = -.02(20)$

$-.02x - .02y = -.4$ We obtained our new first equation.

$-.02x - .02y = -.4$ Add the equations.
$\underline{.02x + .10y = 1.6}$

$.08y = 1.2$ Divide both sides by .08.

$y = 15$ This is the number of liters of 10% solution.

Since we know that $y = 15$, we use the first equation to solve for y.

$x + 15 = 20$

$x = 5$ This is the number of liters of the 2% solution.

Answer: We need 5 liters of the 2% solution and 15 liters of the 10% solution.

Check the answer:

$5 + 15 = 20$ The number of liters checks out.

$5(.02) + 15(.1) = 1.6$ The percents check out also.

Practice Problems

Solve the following.

1. The sum of two numbers is −15. Their difference is 25. Find the two numbers.
2. The sum of two numbers is −10. Their difference is 4. Find the two numbers.
3. Jerry invested $50,000 and earned $3,100 in interest after one year. If some of the money earned 5%, and the remainder earned 8%, then how much did he invest in each account?
4. Greg invested $70,000 and earned $5,800 in interest after one year. If some of the money earned 4%, and the remainder earned 10%, then how much did he invest in each account?
5. Amanda went to the grocery store and purchased six apple pies and three half-gallons of ice cream for $36.00. The next day Marcie went back and purchased four more of the same apple pies and one more half-gallon of ice cream for $21.00. How much was Marcie spending for each apple pie and half-gallon of ice cream?
6. John is trying to create a 20 pound mixture of peanuts and cashews that sell for $5.00 a pound. If peanuts are $2.75 a pound, and cashews are $6.00 a pound, how much of each will he need?
7. A chemist wants to mix a 4% acidic solution with a 10% acidic solution in order to obtain 15 liters of a 6% acidic solution. How much of each solution will be needed in order to obtain this mixture?
8. A chemist wants to mix a 25% acidic solution with water in order to obtain 30 liters of a 10% acidic solution. How much of each will be needed in order to obtain this mixture? (Hint: Water is 0% acid.)

Section 3
Graphing Systems of Inequalities

Introduction

We have already discussed graphing systems of equations (Chapter 4, Section 1) and graphing single inequalities (Chapter 3, Section 4). In this section, we will talk about graphing systems of inequalities.

Learning Objectives

When you finish your study of this chapter, you should be able to
- Graph systems of inequalities in two dimensions

Graphing Systems of Inequalities

To graph a system of inequalities, graph all of the inequalities on the same coordinate plane and then shade the area that they ALL have in common. We recommend that you use two--or three if the problem contains three inequalities--different colors to shade the area. Follow this list of steps for graphing these systems.

Step 1: Graph both lines by finding the x-intercepts and y-intercepts. (A line may also be horizontal or vertical.) Pay attention to whether you should draw a solid or dotted line.

Step 2: Use the test point $(0,0)$ to determine which side of each line to shade. (If $(0,0)$ is a point on the line, use some other point as your test point.)

Step 3: Shade with your two colors. Your answer is the area shaded in both colors.

Let's look at a relatively simple example.

Example 1
Graph: $x \leq 3$
$y \geq 2$

Step 1: First, recall that $x = 3$ is a vertical line and $y = 2$ is a horizontal line. To graph $x \leq 3$, we draw a solid--indicated by the equal sign--vertical line through 3 on the x-axis. To graph $y \geq 2$, draw a solid--again, indicated by the equal sign--horizontal line through 2 on the y-axis.

Step 2: Testing $(0,0)$ in the first inequality, we find that $0 \leq 3$ is true, so we shade on the side of the line that does contains $(0,0)$. Testing $(0,0)$ in the second inequality, we find that $0 \geq 2$ is false, so we shade on the side of the line that does **not** contain $(0,0)$.

Step 3: Use one color to shade to the left of the first line, since we are interested in values less than or equal to 3. Use the second color to shade above the second line, since we are interested in values greater than or equal to 2:

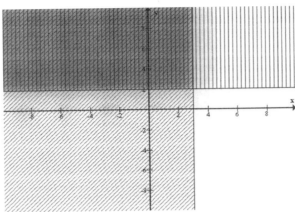

The answer to this problem is the area in the first and second quadrants that is shaded in both colors. (In order to see this shaded area precisely, we recommend using magic markers.)

Example 2

Graph: $y > -3$

$x + y < -1$

Step 1: $y = -3$ is a horizontal line through -3 on the y-axis. Notice that it is dotted. To find the x-intercept for $x + y < -1$: Let $y = 0$. Then, $x + 0 = -1$, so $x = -1$.

The x-intercept is $(-1, 0)$.

To find the y-intercept: Let $x = 0$. Then, $0 + y = -1$, so $y = -1$

The y-intercept is $(0, -1)$.

Notice that the line through these two points is also dotted.

Step 2: We shade above the line $y = -3$ for the first inequality. For the second inequality, the test point $(0,0)$ implies that $0 + 0 < -1$, a sentence which is false. Thus, we shade the side of the line that does **not** contain $(0,0)$.

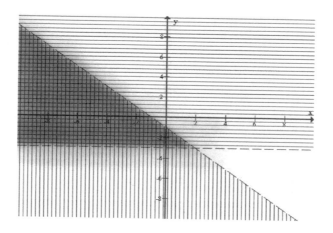

The solution is the darker shaded region.

Example 3

Graph: $x + y < -4$

$3x - 6y > 2$

Step 1: For $x + y = -4$,

To find the x-intercept: Let $y = 0$. Then, $x + 0 = -4$, so $x = -4$. The x-intercept is $(-4, 0)$.

To find the y-intercept: Let $x = 0$. Then, $0 + y = -4$, so $y = -4$. The y-intercept is $(0, -4)$.

Notice that the line through these two points is dotted.

For $3x - 6y = 2$,

To find the x-intercept: Let $y = 0$. Then, $3x - 6(0) = 2$, so $x = 2/3$. The x-intercept is $(2/3, 0)$.

To find the y-intercept: Let $x = 0$. Then, $3(0) - 6y = 2$, so $y = -1/3$. The y-intercept is $(0, -1/3)$.

Notice that the line through these two points is dotted.

Step 2: Testing $(0,0)$ in the first inequality, we find that $0 + 0 < -4$ is false, so we shade on the side of the line that does **not** contain $(0,0)$. Testing $(0,0)$ in the second inequality, we find that $3(0) - 6(0) > 2$ is also false, so, again, we shade on the side of the line that does **not** contain $(0,0)$.

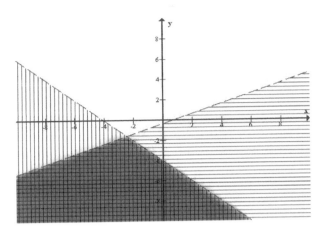

The solution is the darker shaded region at the bottom.

Example 4

Graph: $4x + 8y \leq 12$

$x + 2y \geq -4$

Step 1: For $4x + 8y = 12$,

To find the x-intercept: Let $y = 0$. Then, $4x + 8(0) = 12$, so $x = 3$. The x-intercept is $(3, 0)$.

To find the y-intercept: Let $x = 0$. Then, $4(0) + 8y = 12$, so $y = 1.5$. The y-intercept is $(0, 1.5)$.

Notice that the line through these two points is solid.

For $x + 2y = -4$,

To find the x-intercept: Let $y = 0$. Then, $x + 2(0) = -4$, so $x = -4$. The x-intercept is $(-4, 0)$.

To find the y-intercept: Let $x = 0$. Then, $0 + 2y = -4$, so $y = -2$. The y-intercept is $(0, -2)$.

Notice that the line through these two points is solid.

Step 2: Testing $(0,0)$ in the first inequality, we find that $0 + 0 \leq 12$ is true, so we shade on the side of the line that does contain $(0,0)$. Testing $(0,0)$ in the second inequality, we find that $0 + 0 \geq -4$ is also true, so we again shade on the side of the line that does contain $(0,0)$.

Step 3:

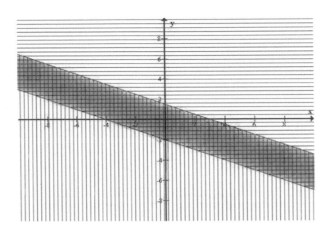

The two lines are parallel, and the solution is the region between the two parallel lines.

Example 5

Graph: $-3x + y < 6$

$y \geq \frac{1}{2}x - 4$

$x \leq 0$

Don't panic just because there are three inequalities. We will still follow the same three steps; we will just look at three equations in each step this time. You may need three magic markers for this problem.

Step 1: For $-3x + y = 6$,

To find the x-intercept: Let $y = 0$. Then, $-3x + 0 = 6$, so $x = -2$. The x-intercept is $(-2, 0)$.

To find the y-intercept: Let $x = 0$. Then, $-3(0) + y = 6$, so $y = 6$. The y-intercept is $(0, 6)$.

Notice that the line through these two points is dotted.

For $y = \frac{1}{2}x - 4$,

To find the x-intercept: Let $y = 0$. Then, $0 = \frac{1}{2}x - 4$, so $x = 8$. The x-intercept is $(8, 0)$.

To find the y-intercept: Let $x = 0$. Then, $y = \frac{1}{2}(0) - 04$, so $y = -4$. The y-intercept is $(0, -4)$.

Notice that the line through these two points is solid.

$x = 0$ is the equation of the y-axis. (Similarly, the equation of the x-axis is $y = 0$.) We draw the y-axis as a solid line.

Step 2: Testing $(0, 0)$ in the first inequality, we find that $-3(0) + 0 < 6$ is true, so we shade on the side of the line that does contain $(0, 0)$. Testing $(0, 0)$ in the second inequality, we find that $0 \geq \frac{1}{2}(0) - 4$ is also true, so we again shade on the side of the line that does contain $(0, 0)$. Finally, we shade to the left of the y-axis, since we are interested in values less than or equal to 0.

Step 3:

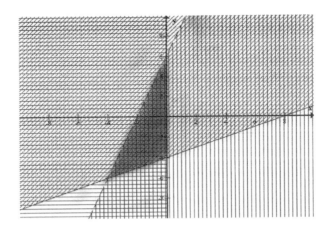

Our final answer is that triangular region in the middle that is shaded the darkest.

Practice Problems

Graph the solution to the systems of inequalities below.

1. $y \geq 0$
   $x \leq 4$

2. $x \geq 0$
   $y \leq -4$

3. $x + 3y < 6$
   $y < 2$

4. $x - y > 3$
   $y > -2$

5. $2x - y \geq 4$
   $3x + 2y \leq 6$

6. $x - 5y \leq 8$
   $4x + 3y \geq 12$

7. $2x + y > 6$
   $4x + 2y > -6$

8. $y < -2$
   $y > 4$

9. $-2x + y \leq 4$
   $y > \frac{1}{2}x + 3$
   $y < 0$

10. $y \leq \frac{-1}{3}x + 3$
    $y \geq \frac{-1}{3}x - 2$
    $y \geq 0$

Chapter 5
Exponents and Polynomials

Assignment Checklist

What You Should Do	Where?			When?	
Read, view the videos, and then complete the online work for Chapter 5, Section 1	📖	💻	MathXL	After completing Chapter 4	
Read, view the videos, and then complete the online work for Section 2	📖	💻	MathXL	After completing Chapter 5, Section 1	
Read, view the videos, and then complete the online work for Section 3	📖	💻	MathXL	After completing Section 2	
Read, view the videos, and then complete the online work for Section 4	📖	💻	MathXL	After completing Section 3	
Read, view the videos, and then complete the online work for Section 5	📖	💻	MathXL	After completing Section 4	
Take the quiz on Chapter 5			MathXL	After completing Section 5	
Post questions and respond to other students' questions in the Discussion Board		💻		Anytime	
Other assignments:					
Notes:					

Section 1
Exponent Laws and Scientific Notation

Learning Objectives

When you finish the study of this section, you should be able to
- Use the product rule to simplify exponential expressions
- Use the zero exponent rule to simplify exponential expressions
- Use the quotient rule to simplify exponential expressions
- Use the negative exponent rule to simplify exponential expressions
- Use the power rule to simplify exponential expressions
- Use scientific notation to write numbers
- Determine the amount of compound interest that accrues on a specific balance

Introduction of Exponents

Remember that an exponent is a number placed as a superscript to the right of a quantity. Given the expression b^n, b is called the base and n is called the exponent. The expression is read as "b to the nth power." Let's review some common introductory problems.

Example 1

Simplify the following.

a. 4^3

$4^3 = 4 \cdot 4 \cdot 4$ The exponent, 3, tells us to multiply three 4s together.

64

Answer: 64

b. -3^2

$-3^2 = -(3 \cdot 3)$ Since the negative sign is not in parentheses, it is not squared with the 3.

$-(9)$ Thus, the answer is -9.

Answer: -9

c. $(-5)^4$

$(-5)^4 = (-5)(-5)(-5)(-5)$ This time the negative sign is inside the parentheses, so it is raised to the fourth power along with the 5.

625 Since the power is even, the answer will be positive.

Answer: 625

d. $(-x)^5$

$(-x)^5 = (-x)(-x)(-x)(-x)(-x)$ The negative is in the parentheses along with the x; thus, both the negative sign and x are raised to the fifth power.

$-x^5$ Since the power is odd, the answer is negative.

Answer: $-x^5$

Exponent Rules

PRODUCT RULE FOR EXPONENTS

If m and n are exponents, and b is a real number, then $b^m \cdot b^n = b^{m+n}$.

In other words, anytime you multiply two quantities with the same base, you keep the base and add the exponents.

Example 2

Simplify $a^2 \cdot a^3$. Since the bases are the same, we add the exponents.

$a^2 \cdot a^3 = a^{2+3}$

a^5

Answer: a^5

Example 3

Simplify $2^3 \cdot 2^5$.

$2^3 \cdot 2^5 = 2^{3+5}$

2^8 (Notice the answer is NOT 4^8)

Answer: 2^8

Example 4

Simplify $(3x^2)(5x^7)$.

$(3x^2)(5x^7) = (3 \cdot 5)(x^2 \cdot x^7)$

$15x^{2+7}$

$15x^9$

Answer: $15x^9$

Example 5

Simplify $(-2a^3b)(5ab^4)$.

$(-2a^3b)(5ab^4) = (-2 \cdot 5)(a^3 \cdot a^1 \cdot b^1 \cdot b^4)$

$-10a^{3+1}b^{1+4}$

$-10a^4b^5$

Answer: $-10a^4b^5$

ZERO EXPONENT RULE

If b does not equal zero, then $b^0 = 1$.

Example 6

Simplify 2^0.

$2^0 = 1$

Since the exponent is 0, and the base is not 0, the answer is 1.

Example 7

Simplify $5b^0$.   Notice b alone is being raised to the zero power, not 5.

$5b^0 = 5 \cdot b^0$

$5 \cdot 1$

5

Answer: 5

Example 8

Simplify $(2x - y)^0$.

Answer: 1

Technically, in this problem, we are assuming that $2x - y \neq 0$.

QUOTIENT RULE FOR EXPONENTS

If m and n are exponents, and b is a non-zero real number, then $\dfrac{b^m}{b^n} = b^{m-n}$.

In other words, anytime you divide two quantities with the same base, you keep the base and subtract the exponents; that is, you take the numerator exponent and subtract the denominator exponent from it.

Example 9

Simplify $\dfrac{x^6}{x^4}$

$$\dfrac{x^6}{x^4} = x^{6-4}$$

$$x^2$$

Answer: x^2

Example 10

Simplify $\dfrac{3^8}{3^3}$

$$\dfrac{3^8}{3^3} = 3^{8-3}$$

$$3^5 \qquad\qquad\text{(Notice the answer is NOT } 1^5)$$

Answer: 3^5

Example 11

Simplify $\dfrac{15x^3y^5}{3x^2y}$

$$\dfrac{15x^3y^5}{3x^2y} = \dfrac{15}{3}x^{3-2}y^{5-1}$$

$$5x^1y^4$$

Answer: $5xy^4$

Example 12, part A

Simplify $\dfrac{8a^4b^2}{6a^2b^3}$

$$\dfrac{8a^4b^2}{6a^2b^3} = \dfrac{8}{6}a^{4-2}b^{2-3}$$

$$\dfrac{4}{3}a^2b^{-1}$$

When the exponent in the denominator is larger than the exponent in the numerator, applying the quotient rule results in a negative exponent, as Example 12 shows. This leads us to the following rule:

NEGATIVE EXPONENT RULE

If b is a real number other than 0, and n is a positive number, then $b^{-n} = \dfrac{1}{b^n}$ and $\dfrac{1}{b^{-n}} = b^n$.

In other words, anytime you have a base with a negative exponent, you must move the base to make the exponent positive. Either you move the term from the numerator to the denominator, or you move the term from the denominator to the numerator, depending on where it originates. Keep in mind that your final answer is **not considered to be simplified** until there are no negative exponents. Having said that, we will now complete Example 12:

Example 12, part B

$$\frac{4a^2 b^{-1}}{3} = \frac{4a^2}{3b}$$

Answer: $\dfrac{4a^2}{3b}$

Example 13

Simplify 3^{-2}.

$3^{-2} = \dfrac{1}{3^2}$ Use the negative exponent rule.

$\dfrac{1}{9}$

Answer: $\dfrac{1}{9}$

Example 14

Simplify $\dfrac{2}{x^{-2}}$.

$\dfrac{2}{x^{-2}} = \dfrac{2x^2}{1}$ Use the negative exponent rule.

$2x^2$

Answer: $2x^2$

Example 15

Simplify $\dfrac{3x^{-1}}{-4y^{-3}}$.

$\dfrac{3x^{-1}}{-4y^{-3}} = \dfrac{3y^3}{-4x^1}$ We move only negative **exponents**; we don't move negative **numbers**.

Answer: $\dfrac{3y^3}{-4x}$ or $\dfrac{-3y^3}{4x}$

Example 16

Simplify $\dfrac{3x^{-5}\,y^3}{12x^{-2}\,y^9}$.

$\dfrac{3x^{-5}\,y^3}{12x^{-2}\,y^9} = \dfrac{3x^2 y^3}{12x^5 y^9}$ Rewrite the problem, first using the negative exponent rule. Then simplify.

$\dfrac{3x^{2-5}\,y^{3-9}}{12}$ Use the quotient rule for exponents.

$\dfrac{3x^{-3}\,y^{-6}}{12}$ Use the negative exponent rule. Cancel.

$\dfrac{1}{4x^3 y^6}$

Answer: $\dfrac{1}{4x^3 y^6}$

POWER RULE FOR EXPONENTS

Sometimes it is necessary to raise exponential expressions themselves to a power. For example:

$$(x^2)^3 = x^2 \cdot x^2 \cdot x^2 = x^{2+2+2} = x^6$$

Notice that if we had multiplied the exponents, the result would be the same:

$$(x^2)^3 = x^{2 \cdot 3} = x^6$$

If m and n are exponents, and b is a real number, then

$$(b^m)^n = b^{mn}$$

In other words, anytime you have a base with a power raised to a power, you keep the base and multiply the exponents.

Example 17

Simplify $(3x^3)^4$.

Since we don't see an exponent, 3 is understood to be raised to the first power.

$(3x^3)^4 = (3)^{1 \cdot 4}(x)^{3 \cdot 4}$

$3^4 x^{12}$

$81x^{12}$

Answer: $81x^{12}$

Example 18

Simplify $(x^2 y^{-3} z)^2$.

$(x^2 y^{-3} z)^2$

$x^{2 \cdot 2} y^{-3 \cdot 2} z^{1 \cdot 2}$

$x^4 y^{-6} z^2$

Using the negative exponent rule to remove the negative exponents gives us the final answer.

Answer: $\dfrac{x^4 z^2}{y^6}$

Example 19

Simplify $(4x^{-3} y^4)^{-3}$.

$(4x^{-3} y^4)^{-3}$

$4^{(1 \cdot -3)} x^{(-3 \cdot -3)} y^{(4 \cdot -3)}$

$4^{-3} x^9 y^{-12}$

Using the negative exponent rule to remove the negative exponents gives us

$\dfrac{x^9}{4^3 y^{12}}$

Answer: $\dfrac{x^9}{64 y^{12}}$

Summary of Exponent Rules

Rule	Notation
Product Rule	$b^m \cdot b^n = b^{m+n}$
Power of Zero	If b does not equal zero, then $b^0 = 1$
Quotient	$\dfrac{b^m}{b^n} = b^{m-n}$
Negative Exponent	$b^{-n} = \dfrac{1}{b^n}$ and $\dfrac{1}{b^{-n}} = b^n$
Power	$(b^m)^n = b^{mn}$

Scientific Notation

Scientific notation is used in science and mathematics in order to express very large or small numbers in a more compact form. For instance, the number $11,900,000,000,000$ (which is the current U. S. National Debt according to the U. S. National Debt Clock) is normally not written like that in a book. Instead, you will see 11.9×10^{13}. This compact form makes writing such numbers much more space friendly for the purposes of formal reports and papers.

Thus, what is the formal definition of scientific notation? Scientific notation means that the number is written in the form $N \times 10^a$, where N must be at least 1 but less than 10, and a must be an integer. Some examples of how to convert numbers to and from scientific notation follow. Notice that when the exponent, a, is positive, the decimal point is moved to the right. When the exponent, a, is negative, the decimal point is moved to the left.

Example 20

Write the following number in scientific notation: $3,400$

3.4×10^3

To increase the value of 3.4 to 3400, the decimal point was moved three places to the **right**.

3.4 to 3400.   Thus, the exponent is **positive**.

Answer: 3.4×10^3

Example 21

Write the following number in scientific notation: .0000567

5.67×10^{-5}

To decrease the value of 5.67 to .0000567, the decimal point was moved five places to the **left**.

5.67 to .0000567   Thus, the exponent is **negative**.

Answer: 5.67×10^{-5}

Example 22

Write the following in standard form: 3.78×10^{-3}

$= .00378$ Move the decimal point three places to the left.

Answer: .00378

Example 23

Write the following in standard form: 1.254×10^6

$= 1,254,000$ Move the decimal point six places to the right.

Answer: 1,254,000

Multiplying and Dividing in Scientific Notation

Scientific notation can speed up calculations with very large or very small numbers. Remember to multiply (or divide) the numerical parts and to add (or subtract) the exponents. Finally, rewrite your final answer in scientific notation, if necessary. Let's look at some examples.

Example 24

Write the product of the following in scientific notation: $(3.45 \times 10^4)(2.1 \times 10^3)$.

First, multiply $3.45 \cdot 2.1 = 7.245$.

Then, use the product rule for the 10s. $10^4 \cdot 10^3 = 10^{4+3}$, or 10^7.

Now, combine the two products above to get the final answer.

Answer: $7.245 \cdot 10^7$

Note: If the product of the first multiplication was not a number between 1 and 10, then we would have had to rewrite the final answer in scientific notation.

Example 25

Write the following quotient in scientific notation: $\dfrac{2.5704 \times 10^{-5}}{1.02 \times 10^2}$.

This problem is equivalent to $\dfrac{2.5704}{1.02} \cdot 10^{-5-2} = 2.52 \cdot 10^{-7}$.

Answer: 2.52×10^{-7}

Applications of Exponents: Compound Interest

For some of you, planning for retirement has already begun. When you plan for retirement, you normally invest your money into accounts that compound the interest on the money. Simply stated, accounts that pay compound interest are paying you money on the interest you've earned as well as the principal you have invested. The formula for compound interest is the following:

$$A(t) = P\left(1 + \dfrac{r}{n}\right)^{nt},$$

where $A(t)$ is the amount in the account after applying the compound interest, r is the rate (in decimal form), n is how often the interest is compounded, and t is the time in years.

If the interest is compounded annually, $n = 1$; semiannually, $n = 2$; quarterly, $n = 4$; monthly, $n = 12$; and daily, $n = 365$.

Let's use this formula in a few problems to see how it works.

Example 26

If you deposited $4,000.00 in an account that compounds interest quarterly at an 8% interest rate, how much would you have after 30 years?

In this problem, $P = 4000$, $r = .08$, $n = 4$, and $t = 30$.

$A(t) = 4000\left(1 + \dfrac{.08}{4}\right)^{(4 \cdot 30)}$ Simplify the exponent and inside the parentheses.

$A(t) = 4000(1 + .02)^{120}$

$A(t) = 4000(1.02)^{120}$

Answer: $A(t) = \$43,060.65$

Example 27

Suppose you learn that in order to retire in 35 years you will need $1,500,000 in an IRA (Individual Retirement Account) to live comfortably. How much of an initial deposit would you need to make (assuming you will add nothing else over the 35 years) in order to obtain this amount? Assume you are earning 10%, and the interest is compounded monthly?

In this problem, $A(t) = 1,500,000$, we do not know $P, r = 0.10$, and $t = 35$. We need to solve the following equation.

$1500000 = P\left(1 + \frac{.10}{12}\right)^{(12 \cdot 35)}$ Simplify the exponent and inside the parentheses.

$1500000 = P(1 + .008333\overline{3})^{420}$ Use a calculator.

$1500000 = P(32.63865043)$ Divide both sides by 32.63865043.

Answer: $\$45,957.78 = P$

Note: We know this is highly unrealistic that you would just make a one-time contribution of this amount into an IRA; however, if you ever came across a good inheritance or signing bonus, this contribution might be feasible.

Practice Problems

Simplify the following.

1. $x^3 \cdot x^4$
2. $x \cdot x^8$
3. $5^2 \cdot 5^4$
4. $3^2 \cdot 3^3$
5. $-4x^2 \cdot (-3x^7)$
6. $-2x \cdot 5x^3$
7. $xy^2 \cdot x^4y^2$
8. $x^0y^3z \cdot x^4y^3z$
9. $(3xy)(-4x^2)$
10. $(-2xy^0)(-8x^4y^2)$
11. $\dfrac{x^5}{x^2}$
12. $\dfrac{6x^5}{10x^2}$
13. $\dfrac{5xy^5}{20x^0y^3}$

14. $\dfrac{10x^4y}{15xy^6}$

15. 3^{-2}

16. a^{-4}

17. $\dfrac{1}{x^{-4}}$

18. $\dfrac{x^{-4}}{y^{-2}}$

19. $\dfrac{3x^{-2}y}{6}$

20. $-5x^{-1}y^2z^{-3}$

21. $\dfrac{x^5 y^{-1}}{x^2 y^3}$

22. $\dfrac{6x^4y^0}{10x^{-2}y}$

23. $\dfrac{3x^{-3}}{12x^4}$

24. $\dfrac{5x^5y^{-4}z^0}{10x^3y^{-1}}$

25. $(x^4)^2$

26. $\left(\dfrac{1}{x^{-3}}\right)^{-3}$

27. $(-3x^2)^3$

28. $(-2x^5y^0)^4$

29. $(4x^{-3}y^4)^3$

30. $(5x^{-2}y^3)^{-2}$

31. $(8x^{-1}yz^4)^{-2}$

32. $(4x^5y^{-3})^{-2}$

33. $\left(\dfrac{5x^4y^6}{15x^{-1}y^9}\right)^{-1}$

34. $(3x^2y^0)(4x^{-1}y^5)$

35. $(5x^4y)(-3x^{-4}y^{-2})$

36. $\dfrac{(2x^{-1}y^2)^3}{16xy^8}$

37. $\left(\dfrac{3x^4}{4y^{-2}}\right)^3$

38. $\left(\dfrac{4xy^3}{5x^5y^{-2}}\right)^2$

39. $\left(\dfrac{4x^2y}{12x^{-4}y^{-2}}\right)^{-3}$

Answer the following.

40. Write as a number: 3.4×10^5.

41. Write as a number: 2.35×10^{-7}.

42. Write in scientific notation: $.0349$.

43. Write in scientific notation: $4,560,000$.

Write the following products and quotients in scientific notation.

44. $(4.2 \times 10^6)(4.78 \times 10^2)$

45. $(6.003 \times 10^{-3})(1.45 \times 10^2)$

46. $(9.03 \times 10^5) \div (3.3 \times 10^{-5})$

47. $(5.065 \times 10^3) \div (1.35 \times 10^2)$

Application problems:

48. If Mary deposits $5,000 in an account that earns 4% compounded quarterly, then how much will she have at the end of 5 years?

49. If Abbey deposits $7,500 in an account that earns 3% compounded monthly, then how much will she have at the end of 10 years?

50. If Rex wants to have $10,000 saved in 7 years, how much of an initial deposit would he need to make in an account that earns 5% compounded monthly?

51. If Bubba needs $1,000,000 in 30 years for retirement, how much of an initial deposit would he need to make in an account that earns 10% compounded quarterly?

Section 2
Polynomials

Learning Objectives

When you finish your study of this section, you should be able to
- Classify a polynomial
- Identify the degree of a polynomial
- Add polynomials
- Subtract polynomials
- Multiply polynomials by using the distributive property
- Multiply polynomials by using the FOIL method

Introduction of Polynomials

Polynomial is a Greek word meaning **many terms**. A polynomial is classified by its number of terms. Terms are separated by addition and subtraction signs. Therefore, if an expression does not have any addition or subtraction signs, it is a **monomial**. If it has two terms, it is a **binomial**, and if it has three terms, it is a **trinomial**. Any polynomial with more than three terms is classified by the general term **polynomial**. The **degree of a monomial in one variable** is the exponent of the variable. The **degree of a monomial in two or more variables** is the **sum** of the exponents of the variables. The **degree of a polynomial** is the highest degree of all of the terms in the polynomial. Let's look at some polynomials in the table below and see how they are classified:

Polynomial	Classification	Degree
$4x^3 - 7$	Binomial	three (exponent on the first term)
$5y^2 + 3y + 2$	Trinomial	two (exponent on the first term)
$x^4y - 5x^3y^3$	Binomial	six (sum of the exponents on the second term)
$x^2y - 4y + 3y^2 - y^3$	Polynomial	three (sum of the exponents on the first term)

Addition of Polynomials

The procedure for adding polynomials follows:

Step 1: Remove the parentheses around each polynomial.

Step 2: Combine like terms.

Once again, two terms are said to be **like** when the terms contain the **same variables with the same exponents**.

Example 1
Add: $(x^2 - 3x + 4) + (3x^2 + 5x + 9)$

$x^2 - 3x + 4 + 3x^2 + 5x + 9$ Remove the parentheses.

$(1+3)x^2 + (-3+5)x + (4+9)$ Combine like terms.

Answer: $4x^2 + 2x + 13$

Example 2
Add: $(3x^3 - 16x + 5) + (7x^3 - 4x^2 + 4x - 2)$

$3x^3 - 16x + 5 + 7x^3 - 4x^2 + 4x - 2$ Remove the parentheses.

Answer: $10x^3 - 4x^2 - 12x + 3$ Combine like terms.

Some instructors prefer the following approach:

Example 3 (Alternative Approach—Vertical addition)
Add: $(3x^3 - 2x^2 + 7x - 10) + (-4x^3 - 10x + 11)$

Step 1: Align like terms and combine them as before.

$$\begin{array}{r} 3x^3 - 2x^2 + 7x - 10 \\ +(-4x^3 + 0x^2 - 10x + 11) \\ \hline -x^3 - 2x^2 - 3x + 1 \end{array}$$

Answer: $-x^3 - 2x^2 - 3x + 1$

Subtraction of Polynomials

The procedure for subtracting polynomials follows:

Step 1: Remove the parentheses from the first polynomial.

Step 2: Remove the parentheses from the second polynomial and write the **opposite** sign **before** each term.

Step 3: Combine like terms.

Example 4

Subtract: $(6x^2 - 5x + 7) - (3x^2 - 5x - 1)$

$6x^2 - 5x + 7 - (3x^2 - 5x - 1)$	Remove first parentheses. Use distributive property.
$6x^2 - 5x + 7 - 3x^2 + 5x + 1$	Combine like terms.
Answer: $3x^2 + 8$	(Notice the answer is **not** $3x^2 + 0x + 8$.)

Example 5

Subtract: $(2x^3 + 12x^2 - 7x - 6) - (-4x^3 + x^2 - 10x - 1)$

$2x^3 + 12x^2 - 7x - 6 - (-4x^3 + x^2 - 10x - 1)$	Use distributive property.
$2x^3 + 12x^2 - 7x - 6 + 4x^3 - x^2 + 10x + 1$	Combine like terms.
$6x^3 + 11x^2 + 3x - 5$	

Answer: $6x^3 + 11x^2 + 3x - 5$

Some instructors prefer the following approach:

Example 6 (Alternative Approach—Vertical Subtraction)

Subtract: $(3x^2 - 5x + 6) - (-x^2 - 7x - 9)$

$$\begin{array}{r} 3x^2 - 5x + 6 \\ -(-x^2 - 7x - 9) \\ \hline \end{array}$$

Change the signs of the second term:

$$\begin{array}{r} 3x^2 - 5x + 6 \\ +\ x^2 + 7x + 9 \\ \hline 4x^2 + 2x + 15 \end{array}$$

Answer: $4x^2 + 2x + 15$

Multiplication of Polynomials

The procedure for multiplying polynomials follows:

Step 1: Use the distributive property to multiply each term in the first polynomial by each term in the second polynomial.

Step 2: Combine like terms, if possible.

Example 7
Multiply: $2x(x-5)$

Step 1: Multiply each term in the first polynomial by each term in the second polynomial.

$2x(x) + 2x(-5)$

Answer: $2x^2 - 10x$

Example 8
Multiply: $-3x^2(x^2 + 4x - 7)$

Step 1: Multiply each term in the first polynomial by each term in the second polynomial.

$-3x^2(x^2) - 3x^2(4x) - 3x^2(-7)$

Answer: $-3x^4 - 12x^3 + 21x^2$

Example 9
Multiply: $(x+4)(x-1)$

Step 1: Multiply each term in the first polynomial by each term in the second polynomial.

$x(x) + x(-1) + 4(x) + 4(-1)$

$x^2 - x + 4x - 4$

Step 2: Combine like terms.

Answer: $x^2 + 3x - 4$

Example 10
Multiply: $(2x-3)(x^2 - 5x + 7)$

Step 1: Multiply each term in the first polynomial by each term in the second polynomial.

$2x(x^2) + 2x(-5x) + 2x(7) - 3(x^2) - 3(-5x) - 3(7)$

$2x^3 - 10x^2 + 14x - 3x^2 + 15x - 21$

Step 2: Combine like terms.

Answer: $2x^3 - 13x^2 + 29x - 21$

Example 11
Multiply: $(x - 2)^3$

$(x - 2)(x - 2)(x - 2)$

$[x(x) + x(-2) - 2(x) - 2(-2)](x - 2)$ Multiply the first two binomials together.

$[x^2 - 2x - 2x + 4](x - 2)$ Combine like terms.

$(x^2 - 4x + 4)(x - 2)$ or $(x - 2)(x^2 - 4x + 4)$ Now multiply by the second binomial.

$x(x^2) + x(-4x) + x(4) - 2(x^2) - 2(-4x) - 2(4)$

$x^3 - 4x^2 + 4x - 2x^2 + 8x - 8$

$x^3 - 6x^2 + 12x - 8$

THE FOIL METHOD

There is a short-cut, called FOIL, to multiplying two binomials.

F – Multiply the first terms in each parentheses.
O – Multiply the two outside terms.
I – Multiply the two inside terms.
L – Multiply the last term in each parentheses.

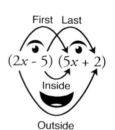

Example 12 (FOIL method)
Multiply: $(2x - 5)(5x + 2)$

$\ \ \ \ \ \text{F}\ \ \ \ \ \ \ \ \text{O}\ \ \ \ \ \ \ \ \ \text{I}\ \ \ \ \ \ \ \ \ \text{L}$
$= 2x(5x) + 2x(2) - 5(5x) - 5(2)$

$10x^2 + 4x - 25x - 10$

$10x^2 - 21x - 10$

Example 13 (FOIL method)
Multiply: $(4x + 1)^2$ (Hint: Recall the definition of an exponent.)

$(4x + 1)(4x + 1)$ Multiply.

$\ \ \ \ \ \text{F}\ \ \ \ \ \ \ \ \text{O}\ \ \ \ \ \ \ \ \ \text{I}\ \ \ \ \ \ \ \ \ \text{L}$
$= 4x(4x) + 4x(1) + 1(4x) + 1(1)$

$16x^2 + 4x + 4x + 1$

Answer: $16x^2 + 8x + 1$

Example 14

Multiply: $(3x - 2y)(3x + 2y)$

$3x(3x) + 3x(2y) - 2y(3x) - 2y(2y)$

$9x^2 + 6xy - 6xy - 4y^2$

$9x^2 - 4y^2$

Answer: $9x^2 - 4y^2$

Practice Problems

Classify each of the following polynomials as a monomial, binomial, or trinomial. Then, state the degree of each.

1. $3x - 6x^2$
2. $5x^3$
3. $2xy - 4y^2 + 3x$
4. $6x^2 - 4x + 1$
5. $9x^2 - 1$
6. 6
7. $5x^4y^3 - 3xy^5$
8. $6x^2y^3z - 3x^2yz^3 + 4x^6y$

Simplify the following:

9. $(3x + 7) + (4x - 5)$
10. $(x^2 - 5x + 8) + (2x^2 - 3x + 7)$
11. $(2x^2 - 3xy + 4y^2) + (6x - xy + y^2)$
12. $(2x^3 - 4x + 7) + (x^2 - 4x - 8)$
13. $(5x^2 - 5x + 1) - (3x^2 + 2x - 4)$
14. $(x^3 - 4x + 5) - (4x^2 - 7x)$
15. $(x^3 - 2x^2 - x - 8) - (3x^3 - 5x + 7)$
16. $(2x^2 - 5xy + 4y^2) - (x^2 - xy + 5y^2)$

Chapter 5: Exponents and Polynomials

17. $4x(2x^2 - 3x + 9)$
18. $-3x^2(4x^3y - 7xy^2)$
19. $5xy(3x - 7y)$
20. $x^2y^2(5xy - 7y^3)$
21. $(x - 3)(x + 4)$
22. $(y - 2)(y - 6)$
23. $(4x - 5)(2x + 1)$
24. $(3a + 4b)(3x - 4y)$
25. $(4a - 6)(5b - 7)$
26. $(4x + 3y)(4x - 3y)$
27. $(3x + 7)(3x - 7)$
28. $(5x + 3y)(3x + 2y)$
29. $(2x^2 + 3)(2x^3 - 3)$
30. $(x^3 - y^3)(x^3 + y^3)$
31. $(6x - 7)^2$
32. $(3x + 5)^2$
33. $(x - 3)^3$
34. $(2x - 5)^3$
35. $(x - 1)^4$
36. $(2x + 1)^4$
37. $(x - 2)(x^2 - 4x + 3)$
38. $(x + 1)(3x^2 - 7x + 10)$
39. $(2x - 1)(x^2 - 2x + 3)$
40. $(3x + 2)(3x^2 - 4x + 1)$
41. $(x^2 - x + 3)(2x^2 - x + 4)$
42. $(y^2 - 3y + 4)(y^2 - y + 2)$

Section 3
Factoring Strategies Part 1

Learning Objectives

When you finish your study of this section, you should be able to
- Factor out a greatest common factor
- Factor using the grouping method
- Factor a difference of squares
- Factor a sum or difference of cubes

Factoring

Writing a polynomial as the product of two or more simpler polynomials is called factoring. For example, since the product of $(x - 4)$ and $(x + 6)$ is $x^2 + 2x - 24$, we say that $x^2 + 2x - 24$ can be factored as the product $(x - 4)(x + 6)$.

This section will focus on some of the common methods used to factor polynomials. The following subsection will complete the discussion of all the factoring methods used in this course. Let's look at the first basic rule of factoring.

Finding the Greatest Common Factor (GCF)

Finding the greatest common factor involves finding a common number and/or variable that is present in every term of the polynomial and then factoring it out. Some helpful steps in completing this process follow.

Step 1: Find the greatest number that divides into all of the numbers without a remainder.
For the example in the first row of the chart on the next page, $3a^3 + 2a^2$, the answer is 1. Therefore, our numerical part of the GCF would be 1.

Step 2: Write the variable(s) that is/are common to all the terms as part of the GCF. Take out the variable with the smaller power.

For the example in the first row of the chart on the next page, is there a variable which is part of each term? The answer is yes. Since the variable a is part of each term, write the variable with the smaller of its two exponents as part of the GCF. The GCF is the product of these two steps. In this case, $1 \cdot a^2 = a^2$. Our GCF is a^2. Other examples follow the chart on the next page.

Polynomial	Numerical Part of GCF	Variable Part of GCF	GCF
$3a^3 + 2a^2$	1	a^2	a^2
$8x + 12$	4	(None)	4
$x^2 + 5x + 6$	1	(None)	1
$12a^3b^4 + 3a^2b^2x^2 - 9ab^3x$	3	ab^2	$3ab^2$

Example 1

Factor: $8x + 12$.

Step 1: Find the greatest number that divides into 8 and without a remainder.

Although 2 divides into 8 and 12 without a remainder, it is not the **largest** number that does so. Therefore, the numerical part of the GCF is 4.

Step 2: Find the variable part of the GCF. Is x part of both terms? The answer is no. Therefore, you do not have a variable part for this GCF. Multiply steps 1 and 2 together to obtain a GCF of 4.

Now, place the GCF in front of the parentheses.

$4(\quad)$

Then, ask yourself: "What do I multiply by 4 to obtain $8x$?"

Write down the answer, "$2x$."

Next, ask yourself: "What do I multiply by 4 to obtain $+12$?"

Write down the answer, "$+3$"

Answer: $4(2x + 3)$

To check your factoring, just use the distributive property. Since $4(2x + 3) = 8x + 12$ your answer is correct.

Example 2

Factor: $12a^3b^4 + 3a^2b^2x^2 - 9ab^3x$.

From the chart just before Example 1, we see the GCF is $3ab^2$. Place the GCF in front of the parentheses.

$3ab^2(\quad)$

Now, ask yourself: "What do I multiply by $3ab^2$ to obtain $12a^3b^4$?"

Write down the answer, "$4a^2b^2$."

Next, ask yourself: "What do I multiply by $3ab^2$ to obtain $+3a^2b^2x^2$?"

Write down the answer, "$+ax^2$."

Finally, ask yourself, "What do I multiply $3ab^2$ by to obtain $-9ab^3x$?"

Write down the answer, "$-3bx$."

Answer: $3ab^2(4a^2b^2 + ax^2 - 3bx)$

Example 3

Factor: $xy + 4y$.

GCF is y.

Ask yourself, "What do I multiply by y to obtain xy?"

Write down the answer, "x."

Finally, ask yourself, "What do I multiply by y to obtain $+4y$?"

Write down the answer, "$+4$."

Answer: $y(x + 4)$

Example 4

Factor: $x(a + b) + 4(a + b)$.

This example is almost identical to Example 3. The difference is that in this example the GCF is a binomial.

GCF is $(a + b)$.

Place the GCF in front of parentheses.

$(a + b)()$ What do I multiply by $(a + b)$ to obtain $x(a + b)$?

$(a + b)(x)$ What do I multiply by $(a + b)$ to obtain $4(a + b)$?

Answer: $(a + b)(x + 4)$

Factoring by Grouping

If you are asked to factor any polynomial that has four terms, you should always try the factoring by grouping method first after verifying that there is no GCF in the polynomial. The procedure for factoring by grouping follows.

Step 1: Factor the first two terms by talking out the GCF and then factor the last two terms by taking out the GCF. The GCF is **negative** when the first term or the third term is negative.

Step 2: Factor the polynomial by using the common binomial as the GCF.

Example 5

Factor: $ax + bx + 4a + 4b$.

Factor the first two terms by using the GCF: $x(a + b)$.

Factor the last two terms by using the GCF: $4(a + b)$.

Let's combine these two pieces:

$= x(a + b) + 4(a + b)$ Now, factor out the common binomial (see Example 4).

Answer: $(a + b)(x + 4)$

Example 6

Factor: $bc + 5b - 3c - 15$.

Factor the first two terms by using the GCF: $b(c + 5)$.

Factor the last two terms by using the GCF: $-3(c + 5)$.

Combine:

$b(c + 5) - 3(c + 5)$ Now, factor out the common binomial.

Answer: $(c + 5)(b - 3)$

Factoring Binomials

When a polynomial has only two terms, it is called a binomial. Whenever you see a binomial, first check to see if there is a GCF. If not, use one of these three methods to factor the binomial.

1. For the difference of two squares, $a^2 - b^2 = (a - b)(a + b)$.
2. For the difference of two cubes, $a^3 - b^3 = (a - b)(a^2 + ab + b^2)$.
3. For the sum of two cubes, $a^3 + b^3 = (a + b)(a^2 - ab + b^2)$.

FACTORING BINOMIALS OF THE FORM $a^2 - b^2$ (DIFFERENCE OF SQUARES)

To recognize a binomial of the form $a^2 - b^2$ (difference of squares), look for the following characteristics:

- It has only two terms
- The two terms are subtracted
- Perfect squares appear: $1, 4, 9, 16, 25, 36, 49, 64, 81, 100, ...$
- The exponents are even numbers

Example 7

Factor: $x^2 - 25$.

Ask yourself, "What variable do I multiply by itself to obtain x^2?"

Write the answer, x, in the first position of both parentheses:

$(x \quad)(x \quad)$

Ask yourself, "What number do I multiply by itself to obtain 25?"

Write the answer, 5, in the second position in both parentheses.

$(x \quad 5)(x \quad 5)$

Finally, one sign is $+$; one sign is $-$. The order doesn't matter (since you do not want to have a middle term when you multiply).

Answer: $(x + 5)(x - 5)$

Check by multiplication: $(x + 5)(x - 5)$

$x^2 + 5x - 5x - 25$ Notice the x terms add to 0 due to the different signs.

$x^2 - 25$ (Correct)

Example 8

Factor: $36x^2 - 49y^2$.

Ask yourself, "What term do I multiply by itself to obtain $36x^2$?"

Write the answer, $6x$, in the first position of both parentheses.

$(6x \quad)(6x \quad)$

Ask yourself, "What term do I multiply by itself to obtain $49y^2$?"

Write the answer, $7y$, in the second position in both parentheses:

$(6x \quad 7y)(6x \quad 7y)$

Finally, one sign is $+$; one sign is $-$. (The order does not matter.)

Answer: $(6x + 7y)(6x - 7y)$

FACTORING BINOMIALS OF THE FORM $a^3 + b^3$ OR $a^3 - b^3$

For the **difference** of two cubes, $a^3 - b^3 = (a - b)(a^2 + ab + b^2)$.

For the **sum** of two cubes, $a^3 + b^3 = (a + b)(a^2 - ab + b^2)$.

To recognize binomials of the form $a^3 + b^3$ or $a^3 - b^3$, look for the following characteristics:

- There are two terms.
- Perfect cubes appear: 1, 8, 27, 64, 125, 216, 343, 512, 729, 1000, ...
- The exponents are multiples of. 3

Step 1: Identify a and by rewriting the problem as a sum or difference of two quantities cubed. For example $8x^3 + 27$ would be rewritten as $(2x)^3 + (3)^3$ and thus $a = 2x$ and $b = 3$.

Step 2: Substitute a and b in the appropriate formula noted above, using parentheses for each substitution.

Step 3: Use your knowledge of evaluating algebraic expressions with the given values for a and b to simplify the expressions.

Example 9

Factor: $27x^3 - 125y^3$.

Step 1: Write each term of the binomial as a difference of two quantities cubed and identify a and b.

$(3x)^3 - (5y)^3$ (Therefore, $a = 3x$ and $b = 5y$.)

Step 2: Substitute a and b in the formula noted above. We will use the difference of two cubes formula since there is a $-$ sign between $27x^3$ and $125y^3$.

$((3x) - (5y))((3x)^2 + (3x)(5y) + (5y)^2)$

Step 3: Simplify the expression where you can.

Answer: $(3x - 5y)(9x^2 + 15xy + 25y^2)$

Check your answer by multiplication.

$(3x - 5y)(9x^2 + 15xy + 25y^2)$

$3x(9x^2 + 15xy + 25y^2) - 5y(9x^2 + 15xy + 25y^2)$ Multiply each term in the first parentheses by each term. in the second parentheses.

$27x^3 + 45x^2y + 75xy^2 - 45x^2y - 75xy^2 + 125y^3$ Combine like terms.

$27x^3 + 125y^3$ (Correct)

Example 10

Factor: $8x^3 + 1$.

Step 1: Write each term of the binomial as a sum of two quantities cubed and identify a and b.

$(2x)^3 + (1)^3$ (Therefore, $a = 2x$ and $b = 1$.)

Step 2: Substitute a and b in the formula noted above. We will use the sum of two cubes formula since there is a $+$ sign between $8x^3$ and 1.

$((2x) + (1))((2x)^2 - (2x)(1) + (1)^2)$

Step 3: Simplify the expression where you can.

Answer: $(2x + 1)(4x^2 - 2x + 1)$

To remember the order of the signs in a sum or difference of cubes problem, remember the magic word *SOAP*. SOAP stands for same sign as the original problem, opposite sign as the original problem, always positive.

Example:

$8x^3 + 1 = (2x + 1)(4x^2 - 2x + 1)$

- SAME SIGN
- OPPOSITE SIGN
- ALWAYS POSITIVE

Practice Problems

Factor completely. If the expression cannot be factored, write the word "prime," which means the same thing as "cannot be factored." (You should remember this term from Elementary Algebra.)

1. $7x - 21$
2. $5x + 25x^2$
3. $13xy - 26x^2y$
4. $16x^3 - 10x$
5. $8x^3 - 12x^2 - 4x$
6. $20x^3y^2 - 15x^3y + 25x^2y^3$
7. $y(x - 1) + 3(x - 1)$
8. $x(x + 4) - 2(x + 4)$
9. $x^3 - 6x^2 + 2x - 12$
10. $x^3 + 3x^2 - 7x - 21$
11. $ab - by - ca + cy$
12. $yx - 5y + zx - 5z$
13. $x^2 - 9$
14. $4x^2 - 25$
15. $x^2 - y^2$
16. $81x^2 - 49y^2$
17. $16x^2 + 9$
18. $4x^2 + 25$

19. $36x^2 - 1$
20. $4x^2 - 16$
21. $27x^3 - 8$
22. $x^3 - 1$
23. $8x^3 + 27y^3$
24. $64x^3 + 125y^3$

Section 4
Factoring Strategies Part 2

Learning Objectives

When you finish your study of this section, you should be able to factor
- Trinomials of the form $ax^2 \pm bx \pm c$ with $a = 1$
- Trinomials of the form $ax^2 \pm bx \pm c$ with $a \neq 1$
- Polynomials by using the substitution method
- Polynomials that require more than one step

Factoring Trinomials of the Form $ax^2 + bx + c$, $a = 1$

We break these problems up into two cases: the last sign is + or the last sign is −. We will call this last sign the **indicator sign** since it indicates whether we will add or subtract to obtain the middle term.

Indicator Sign Is +

If the problem looks like $x^2 + bx + c$, its factors look like $(x + \underline{})(x + \underline{})$.

If the problem looks like $x^2 - bx + c$, its factors look like $(x - \underline{})(x - \underline{})$.

Notice that when the indicator sign is +, the two signs in the answer are the same; either both are positive or both are negative because we are adding to obtain the middle term.

When the indicator sign is +, we are looking for two numbers that equal c when multiplied together and equal b when added together.

Example 1

Factor: $x^2 + 7x + 10$.

Since the signs are + and +, we know the answer looks like $(x + \underline{})(x + \underline{})$.

What we have to do is figure out the numbers that fill in the blanks.

We are looking for two numbers that **multiply** to 10 and **add** to 7.

The possibilities are 1 & 10 and 2 & 5.

Only the 2 and 5 will multiply to 10 and add to 7, so those are the numbers we will use.

Thus, $x^2 + 7x + 10 = (x + 2)(x + 5)$.

Answer: $(x + 2)(x + 5)$

We check our answer by multiplying or using FOIL:

$x^2 + 5x + 2x + 10$

$x^2 + 7x + 10$

Example 2

Factor: $y^2 - 9y + 18$.

Since the signs are − and +, we know the answer looks like $(x - \_\_\_)(x - \_\_\_)$.

We are looking for two numbers that **multiply** to 18 and **add** to 9.

The possibilities are 1 & 18, 2 & 9, and 3 & 6.

Only 3 and 6 multiply to 18 and add to 9, so those are the numbers we will use.

Thus, $y^2 - 9y + 18 = (y - 3)(y - 6)$.

Answer: $(y - 3)(y - 6)$

Check the answer (use the FOIL method):

If you have ever wondered how a teacher can look at your answer and in a few seconds determine if you are correct, then this will end your puzzlement. What your teacher is doing is multiplying the inner and outer terms together to see if they combine to create the correct middle term. See below:

$(y - 3)(y - 6)$

$-3y$	This is the product of the inner terms.
$-6y$	This is the product of the outer terms.
$-9y$	Combining $-3y$ and $-6y$ gives us $-9y$, which is the middle term of the original equation. If this sum is wrong, we have chosen the wrong factors.

Since it's right, now we check the first and last terms.

y^2	This is the product of the first terms in each parenthesis.
$+18$	This is the product of the last terms in each parentheses.
$y^2 - 9y + 18$	Putting all the parts together will verify the answer. Again, most teachers can do this in their heads; can you? Try it on the next few examples.

Example 3

Factor: $x^2 - 11x + 28$.

Since the signs are $-$ and $+$, we know the answer looks like $(x - \underline{})(x - \underline{})$.

We are looking for two numbers that **multiply** to 28 and **add** to 11.

The possibilities are 1 & 28, 2 & 14, and 4 & 7.

The numbers that work are 4 and 7.

Thus, $x^2 - 11x + 28 = (x - 4)(x - 7)$.

Answer: $(x - 4)(x - 7)$

Try to check this mentally by using the FOIL method mentioned at the end of Example 2.

INDICATOR SIGN IS −

If the problem looks like $x^2 + bx - c$, its factors look like $(x + \underline{})(x - \underline{})$.

If the problem looks like $x^2 - bx - c$, its factors look like $(x + \underline{})(x - \underline{})$.

Notice that when the indicator sign is $-$, the signs in the answer are different; one is positive, and one is negative because we are subtracting to obtain the middle term. The order of the signs does not matter as long as the larger number matches the sign of b.

When the indicator sign is $-$, we are looking for two numbers that equal c when multiplied together and equal b when subtracted. The larger of the two numbers goes with the sign of b, so when the middle term is negative, the larger factor will be negative. If the middle term is positive, then the larger factor will be positive. Let's look at some examples.

Example 4

Factor: $x^2 + 5x - 24$.

Since the signs are $+$ and $-$, its factors look like $(x + \underline{})(x - \underline{})$.

We are looking for numbers that **multiply** to 24 and **subtract** to 5.

The possibilities are 1 & 24, 2 & 12, 3 & 8, and 4 & 6.

The numbers that work are 3 and 8.

The larger number is 8, so it goes with the middle sign in the problem, which is $+$.

Thus, $x^2 + 5x - 24 = (x + 8)(x - 3)$ or $(x - 3)(x + 8)$.

Answer: $(x - 3)(x + 8)$

Example 5

Factor: $y^2 - 5y - 50$.

Since the signs are $-$ and $-$, its factors look like $(x + \underline{})(x - \underline{})$.

We are looking for numbers that **multiply** to 50 and **subtract** to 5.

The possibilities are 1 & 50, 2 & 25, and 5 & 10.

The numbers that work are 5 and 10.

The larger number is 10, so it goes with the middle sign in the problem, which is $-$.

Thus, $y^2 - 5y - 50 = (y + 5)(y - 10)$.

Answer: $(y + 5)(y - 10)$

Example 6

Factor: $x^2 + 7x - 18$.

Since the signs are $+$ and $-$, its factors look like $(x + \_\_\_)(x - \_\_\_)$.

We are looking for numbers that **multiply** to 18 and **subtract** to 7.

The possibilities are 1 & 18, 2 & 9, and 3 & 6.

The numbers that work are 2 and 9.

The larger number is 9, so it goes with the middle sign in the problem, which is $+$.

Thus, $x^2 + 7x - 18 = (x + 9)(x - 2)$.

Answer: $(x + 9)(x - 2)$

Factoring Trinomials of the Form $ax^2 \pm bx \pm c, a \neq 1$

INDICATOR SIGN IS +

These problems are a bit more complicated. However, when the indicator sign (last sign) is $+$, we are still looking for two numbers that **multiply** to $a \cdot c$ and **add** to b. We are going to call this the AC, or grouping, method for factoring these types of trinomials. The steps for using the AC, or grouping, method follow:

Step 1: Multiply a by c.

Step 2: Write out all the factors of the product of ac and choose the factors that combine to bx. Watch the signs!

Step 3: Substitute the factors for the middle term, bx, of the original equation.

Step 4: Factor the new polynomial by grouping.

Step 5: Check the factoring by multiplying or the FOIL method.

Example 7

Factor: $3x^2 + 8x + 4$.

Step 1: Multiply a by c: $3 \times 4 = 12$.

Step 2: The factors of twelve are 1 & 12, 2 & 6, and 3 & 4. You are looking for the factors that **add** to $+8x$. *What are those numbers?* They are $+6x$ and $+2x$.

Step 3: Write bx as the sum of the numbers you just found.

$3x^2 + 8x + 4$ is the original equation.

$3x^2 + 6x + 2x + 4$ is the new equation after substitution. Do not combine terms. The order of the $2x$ and $6x$ does not matter.

Step 4: Now factor the polynomial by grouping:

$(3x^2 + 6x) + (2x + 4)$ Find the GCF for each pair of terms.

$3x(x + 2) + 2(x + 2)$ Factor out the GCF from each pair.

Answer: $(x + 2)(3x + 2)$

Step 5: Check your answer. Use the multiplication method.

$3x^2 + 2x + 6x + 4$

$3x^2 + 8x + 4$ (Correct)

Example 8

Factor: $2x^2 - 9x + 9$.

Step 1: Multiply a by c: $2 \times 9 = 18$.

Step 2: The factors of 18 are 1 & 18, 2 & 9, and 3 & 6. You are looking for the factors that add to $-9x$, so both numbers must be negative. You need -3 & -6 or $-3x$ & $-6x$ for substitution purposes.

Step 3: Write bx as the sum of the numbers you just found.

$2x^2 - 9x + 9$ The original equation

$[2x^2 - 6x] + [-3x + 9]$ The new equation with substitution

Step 4: Factor by grouping:

$2x(x - 3) - 3(x - 3)$ Factor out the GCF from each pair of terms.

Answer: $(x - 3)(2x - 3)$

Step 5: Remember to check your answer. Try checking this one mentally using the FOIL method.

Indicator Sign Is −

When the last sign is −, we are looking for numbers that **multiply** to ac and **subtract** to b. We will use the same steps noted above for the AC, or grouping, method.

Example 9

Factor: $8x^2 - 2x - 3$.

Step 1: Multiply a by c: $8 \times 3 = 24$.

Step 2: The factors of 24 are 1 & 24, 2 & 12, 3 & 8, and 4 & 6. You want the factors that **subtract** to $-2x$. Choose $-6x$ and $+4x$. (Remember that you want $-2x$ in the middle, so the $6x$ term (the **larger** one) must be negative.)

Step 3: Substitute the $+4x$ and $-6x$ for the middle term of $-2x$ in the original equation.

$[8x^2 + 4x] + [-6x - 3]$

Step 4: Factor by grouping:

$4x(2x + 1) - 3(2x + 1)$

Answer: $(2x + 1)(4x - 3)$

Example 10

Factor: $3x^2 + x - 10$.

Step 1: Multiply a by c: $3 \times 10 = 30$.

Step 2: The factors of 30 are 1 & 30, 2 & 15, 3 & 10, and 5 & 6. You are looking for factors that **subtract** to $1x$. Choose $-5x$ and $+6x$.

Step 3: Substitute $+6x$ and $-5x$ for the middle term of $+x$ in the original equation.

$[3x^2 + 6x] + [-5x - 10]$

Step 4: Factor by grouping:

$3x(x + 2) - 5(x + 2)$

Answer: $(x + 2)(3x - 5)$

Example 11 (Trial and Error Method)

Factor: $3x^2 + x - 10$

Step 1: Determine all the factors of the first term, $3x^2$.

The only factors of $3x^2$ are $3x$ and x.

Step 2: Determine all the factors of the last term, 10.

The factors of 10 are 1 & 10 or 2 & 5.

Step 3: Now write out all the possible combinations of factors.

Hint: Fix the $3x$ and x, and just move the factors of 10 around.

$(3x \quad 10)(x \quad 1)$ $(3x \quad 1)(x \quad 10)$

$(3x \quad 2)(x \quad 5)$ $(3x \quad 5)(x \quad 2)$

Step 4: Now use the FOIL method to determine which set of factors will subtract to give you x.

Option	A	B	C	D
Factor Combination	$(3x \quad 10)(x \quad 1)$	$(3x \quad 1)(x \quad 10)$	$(3x \quad 2)(x \quad 5)$	$(3x \quad 5)(x \quad 2)$
Multiply the inner and outer terms:	$10x$ and $3x$	$1x$ and $30x$	$2x$ and $15x$	$5x$ and $6x$
Subtract the two terms:	$7x$	$29x$	$13x$	x

The only option that gives you x is d. Now make sure you choose signs that give you $+x$, not $-x$.

Thus, $3x^2 + x - 10 = (x + 2)(3x - 5)$.

Answer: $(x + 2)(3x - 5)$

Factoring by Substitution

Sometimes we are given polynomials to factor that involve what is called the substitution method. The substitution method involves using a new variable to replace a more complicated expression. Let's look at some examples.

Example 12

Factor: $y^6 - 9$.

We would know how to factor this expression if it was a difference of two squares. Let's try to write it as a difference of two squares. *What term squared would give us y^6?* The answer is y^3.

We let $x = y^3$, and by raising both sides to the power of two, we obtain $x^2 = y^6$.

Thus, we can substitute x^2 for y^6 in the original problem.

By substituting x^2 for y^6, we now have a new expression, $x^2 - 9$.

This expression factors into $(x + 3)(x - 3)$.

Since we substituted $x = y^3$ in the beginning, we need to reverse the substitution at the end to get the final answer.

Answer: $(y^3 + 3)(y^3 - 3)$

Example 13

Factor: $x^{2u} - x^u - 6$.

We've never factored anything with an x^u in it. Let's try letting $q = x^u$.

Then, $q^2 = (x^u)^2 = x^{2u}$.

Next, substitute q for x^u and q^2 for x^{2u}.

Now, we have the trinomial $q^2 - q - 6$.

Using the rules for factoring trinomials gives us the following **answer:** $(q-3)(q+2)$.

Lastly, reversing the substitution of x^u for q gives us the final answer.

Answer: $(x^u - 3)(x^u + 2)$

Multi-Step Factoring

Many times factoring polynomials involves more than one factoring method. Some examples of multi-step problems follow.

Example 14

Factor: $x^3 - 4x$.

Step 1: Factor out the GCF, giving you $x(x^2 - 4)$.

Step 2: Check to see if the expression in the parentheses can be factored again. In this case, $x^2 - 4$ can be factored, by using the difference of two squares rule, into $(x+2)(x-2)$.

Step 3: To write the final answer, remember to include the original x that was factored out.

Answer: $x(x+2)(x-2)$

Example 15

Factor: $3x^2y + 21xy + 36y$.

Step 1: Factor out the GCF, giving you $3y(x^2 + 7x + 12)$.

Step 2: Since $x^2 + 7x + 12$ can be factored into $(x+3)(x+4)$ by using the trinomial factoring methods, your final answer will be the following.

Answer: $3y(x+3)(x+4)$

Example 16

Factor: $x^6 - 1$.

Step 1: If you do not wish to use the substitution method shown above, recognize that this binomial is a difference of two squares: the exponent is even, the two terms are subtracted, and 1 is a perfect square.

$(x^3 - 1)(x^3 + 1)$

Step 2: Both binomials can now be factored using the sum and difference of cubes rules mentioned in the previous section:

$x^3 - 1$ factors into $(x - 1)(x^2 + x + 1)$, and $x^3 + 1$ factors into $(x + 1)(x^2 - x + 1)$, giving you the final answer.

Answer: $(x - 1)(x^2 + x + 1)(x + 1)(x^2 - x + 1)$

General Strategy for Factoring

Step 1: Are there any common factors? If so, try factoring by using the GCF method.

Step 2: Count the number of terms in the polynomial.

1. If there are two terms, then try using one of the following methods:
   a. Difference of two squares: $a^2 - b^2 = (a - b)(a + b)$
   b. Difference of two cubes: $a^3 - b^3 = (a - b)(a^2 + ab + b^2)$
   c. Sum of two cubes: $a^3 + b^3 = (a + b)(a^2 - ab + b^2)$

2. If there are three terms, then try a trinomial method.
   a. Trinomials of the form $ax^2 \pm bx \pm c$, $a = 1$
   b. Trinomials of the form $ax^2 \pm bx \pm c$, $a \neq 1$

3. If the polynomial contains four or more terms, then try factoring by grouping.

Step 3: Check to see if any factors in the factored polynomial can be factored any further. If so, then factor them.

Practice Problems

Factor completely. If the expression cannot be factored, write "prime".

4. $x^2 - 3x - 4$
5. $x^2 + 5x + 6$
6. $x^2 - 2x - 15$
7. $x^2 + 8x + 12$
8. $x^2 - 13x + 36$

9. $x^2 + x - 42$

10. $2x^2 - 7x + 5$

11. $3x^2 - 5x - 2$

12. $3x^2 + 10x + 8$

13. $5x^2 - 31x + 6$

14. $10x^2 - 7x - 6$

15. $15x^2 - 31x + 10$

16. $12x^2 + xy - 6y^2$

17. $16x^2 + 32xy + 15y^2$

18. $x^4 - 9$

19. $x^8 - 25$

Factor the following by using the substitution method.

20. $x^{2u} - 4x^u - 5$

21. $x^{2u} + 7x^u + 12$

22. $y^6 - 4y^3 + 3$

23. $y^8 - y^4 - 2$

24. $x^{2n} - 4$

25. $y^{6n} - 16$

Factor the following by using any and all methods necessary.

26. $y^3 - y$

27. $x^2y - 25y$

28. $x^3 - x^2 - 20x$

29. $xy^2 + 7xy - 20x$

30. $x^2y^3 - 8x^2$

31. $x^6 - 64$

Section 5
Solving Quadratic Equations by Using the Factoring Method

Learning Objectives

When you finish your study of this section, you should be able to
- Solve a quadratic equation by using the factoring method

Points to Remember

General Form of a Quadratic Equation

A quadratic equation is in the form $ax^2 \pm bx \pm c = 0$ where $a \neq 0$, and a, b, and c are real numbers. In other words, the largest exponent of the variable for which you are solving is 2. In this section, we focus on solving quadratic equations by factoring.

Principle of Zero Product

If $a \cdot b = 0$, then either $a = 0$, or $b = 0$, or both a and $b = 0$.

In other words, if two numbers are multiplied together, and the answer is 0, either the first number is 0, the second number is 0, or both numbers are 0.

This principle allows us to solve quadratic equations by factoring.

To Solve a Quadratic Equation by Factoring

Step 1: Write the equation in standard form with the highest degree term positive. This will result in one side of the equation being 0.

Step 2: Factor the non-zero side of the equation.

Step 3: Set each factor containing a variable equal to zero, and solve each equation.

(As always, check each solution found in Step 3 in the original equation.)

Example 1
Solve: $x^2 - 4x = 0$.

Step 1: Write the equation in standard form with the squared term positive. This will result in one side of the equation being zero. This problem is already in standard form:

$x^2 - 4x = 0$.

Step 2: Factor the non-zero side of the equation:

$x(x - 4) = 0$

Step 3: Set each factor containing a variable equal to zero and solve:

$x = 0; x - 4 = 0$

Answers: $x = 0; x = 4$

Check:

Does $0^2 - 4(0) = 0$? Yes.

Does $4^2 - 4(4) = 0$? Yes.

Example 2

Solve: $x^2 - 11x = 60$.

Step 1: Write the equation in standard form with the squared term positive. This will result in one side of the equation being 0. Subtract 60 from both sides:

$x^2 - 11x - 60 = 0$.

Step 2: Factor the non-zero side of the equation:

$(x - 15)(x + 4) = 0$

Step 3: Set each factor containing a variable equal to zero and solve:

$x - 15 = 0; x + 4 = 0$

Answers: $x = 15; x = -4$

Example 3

Solve: $-x^2 + 7x - 12 = 0$.

Step 1: One side already equals zero, but the coefficient of the squared term is -1. Multiply each term by -1 to change the sign of the leading coefficient:

$(-1)(-x^2 + 7x - 12) = (-1)(0)$

$x^2 - 7x + 12 = 0$

Step 2: Factor the non-zero side:

$(x - 4)(x - 3) = 0$

Step 3: Set each factor equal to zero and solve:

$x - 4 = 0; x - 3 = 0$

Answers: $x = 4; x = 3$

Example 4

Solve: $x^2 = 36$.

Step 1: Write the equation in standard form with the squared term positive. This will result in one side of the equation being 0. Subtract 36 from both sides:

$x^2 - 36 = 0$.

Step 2: Factor the non-zero side of the equation:

$(x - 6)(x + 6) = 0$

Step 3: Set each factor containing a variable equal to zero and solve:

$x - 6 = 0; \ x + 6 = 0$

Answers: $x = 6; \ x = -6$

Example 5

Solve: $4x^2 - 7x = 2$.

Step 1: The 2 needs to be brought to the left side of the equation so that the equation is in standard form. In order to do this, subtract 2 from both sides:

$4x^2 - 7x - 2 = 2 - 2$

$4x^2 - 7x - 2 = 0$

Step 2: Now factor the non-zero side of the equation:

$(4x + 1)(x - 2) = 0$

Step 3: Set each factor equal to zero and solve:

$4x + 1 = 0; \ x - 2 = 0$

$4x = -1; \ x = 2$

Answers: $x = \dfrac{-1}{4}; \ x = 2$

Example 6

Solve: $6x^2 = 24$.

$6x^2 - 24 = 0$	Subtract 24 from both sides.
$6(x^2 - 4) = 0$	Factor out the GCF of 6.
$6(x - 2)(x + 2) = 0$	Factor $x^2 - 4$.
$x - 2 = 0; \ x + 2 = 0$	Set each factor equal to 0 and solve.

Answers: $x = 2; \ x = -2$

Notice that we don't set the 6 equal to 0, since there is no variable with the 6.

Example 7

Solve: $x(x-2) = 48$.

First, use the distributive property in order to write the equation in standard form.

$x^2 - 2x = 48$ Subtract 48 from both sides.

$x^2 - 2x - 48 = 0$ Factor.

$(x-8)(x+6) = 0$ Set each factor equal to zero and solve.

$x - 8 = 0;\ x + 6 = 0$

Answers: $x = 8;\ x = -6$

Example 8

Solve: $4x^3 - 16x^2 = 20x$.

$4x^3 - 16x^2 - 20x = 0$ Factor completely.

$4x(x^2 - 4x - 5) = 0$

$4x(x-5)(x+1) = 0$ Set each factor equal to 0 and solve.

$4x = 0;\ x - 5 = 0;\ x + 1 = 0$

Answers: $x = 0;\ x = 5;\ x = -1$

You may notice that this equation isn't a quadratic equation, since the largest exponent of the variable for which you are solving is 3. However, we can still use the methods of this section to solve the equation, so we threw it in as a bonus. Did you notice that when the largest exponent was 3, there were three solutions to the equation? Do you think that's a coincidence?

Example 9

Solve: $\frac{2}{3}x^2 + \frac{1}{3}x - 1 = 0$.

$3\left(\frac{2}{3}x^2\right) + 3\left(\frac{1}{3}x\right) - 3(1) = 0$ Multiply through by 3 to clear the fractions.

$2x^2 + x - 3 = 0$ Factor using the AC, or grouping, method.

$(2x + 3)(x - 1) = 0$

$2x + 3 = 0;\ x - 1 = 0$

$2x = -3;\ x = 1$

Answers: $x = \frac{-3}{2};\ x = 1$

Practice Problems

Solve the following equations for x.

1. $(3x - 2)(x + 4) = 0$
2. $x(x - 1)(4x - 5) = 0$
3. $4(x - 3)(2x + 1) = 0$
4. $2x(3x - 5) = 0$
5. $x^2 - 4x = 0$
6. $3x^2 - 15x = 0$
7. $x^2 - 2x - 24 = 0$
8. $-x^2 - 5x - 6 = 0$
9. $x^2 - 64 = 0$
10. $4x^2 - 49 = 0$
11. $x^2 - x - 20 = 0$
12. $x^2 + 12x + 36 = 0$
13. $x^2 - 7x = -12$
14. $x^2 - 8 = -2x$
15. $4x^2 - x - 3 = 0$
16. $3x^2 + 8x + 4 = 0$
17. $7x^2 + 17x = -6$
18. $12x^2 - 5 = 11x$
19. $2x^2 - 8 = 0$
20. $x^3 - 9x = 0$
21. $3x^2 - 21x + 30 = 0$
22. $4x^2 + 6x - 10 = 0$
23. $x(x - 5) = -4$
24. $(x - 2)^2 = 16$
25. $\frac{1}{3}x^2 - \frac{10}{3}x + \frac{16}{3} = 0$
26. $\frac{5}{2}x^2 - \frac{19}{2}x = -2$

Chapter 6
Rational Expressions and Equations

Assignment Checklist

What You Should Do	Where?			When?	✓
Read, view the videos, and then complete the online work for Chapter 6, Section 1	📖	💻	MathXL	After completing Chapter 5	
Read, view the videos, and then complete the online work for Section 2	📖	💻	MathXL	After completing Chapter 6, Section 1	
Read, view the videos, and then complete the online work for Section 3	📖	💻	MathXL	After completing Section 2	
Read, view the videos, and then complete the online work for Section 4	📖	💻	MathXL	After completing Section 3	
Read, view the videos, and then complete the online work for Section 5	📖	💻	MathXL	After completing Section 4	
Read, view the videos, and then complete the online work for Section 6	📖	💻	MathXL	After completing Section 5	
Read, view the videos, and then complete the online work for Section 7	📖	💻	MathXL	After completing Section 6	
Take the quiz on Chapter 6			MathXL	After completing Section 7	
Take the practice test on Chapters 4-6			MathXL	After completing Chapter 6 quiz	
Schedule your test with your instructor				After completing practice test on Chapters 4-6	
Post and read responses to other students' questions in the Discussion Board		💻		Anytime	
Other assignments:					
Notes:					

Section 1
Rational Expressions

Learning Objectives

When you finish your study of this section, you should be able to
- Recognize rational expressions and rational functions
- Evaluate a rational function when given values for the variable
- Find the domain of a rational function
- Reduce a rational expression to lowest terms

Rational Expressions and Functions

A rational expression is a fraction in which the numerator and denominator are polynomials. In symbols, a rational expression can be written as the quotient $\dfrac{p(x)}{q(x)}$ in which $p(x)$ and $q(x)$ are polynomials.

Examples of Rational Expressions

a. $\dfrac{24x^6y^5}{8x^7y}$
b. $\dfrac{y^2 - 3y + 2}{y^2 - 4y + 3}$
c. $\dfrac{3x^2 - 5x - 2}{6x^3 + 2x^2 + 3x + 1}$

A rational function is a function consisting of a rational expression. However, the denominator of a rational function is not allowed to equal 0, since we can't divide by 0. (Hint: Recall that $3 \div 0$ is undefined.)

Examples of Rational Functions

a. $f(x) = \dfrac{5}{x}$
b. $f(x) = \dfrac{6}{2x - 3}$
c. $f(x) = \dfrac{2x + 1}{x^2 - x - 2}$

Evaluating Rational Functions

We can evaluate rational functions by substituting the value of the variable in the function and performing the indicated operations.

Example 1
Evaluate the rational function $f(x) = \dfrac{x^2 - 3x + 4}{x - 2}$ when $x = 1$

Substitute the number 1 for every variable x. After we substitute 1 for the variable, we simplify the numerator and denominator.

$$\frac{(1)^2 - 3(1) + 4}{(1) - 2}$$

Answer: $\frac{1 - 3 + 4}{-1} = \frac{-2 + 4}{-1} = \frac{2}{-1} = -2$

Example 2

Evaluate the rational function $f(x) = \frac{x + 5}{x^2 - 9}$ when $x = -3$

Substitute the number -3 for every variable x.

$$\frac{(-3) + 5}{(-3)^2 - 9}$$

Answer: $\frac{-3 + 5}{9 - 9} = \frac{2}{0} =$ undefined.

Substituting the number -3 gave us an undefined expression. This demonstrates that we are not allowed to replace x with -3 in this particular rational function. For this rational function, we say $x \neq -3$. It also means that -3 is not part of the domain of the function, since -3 does not have a corresponding function value.

FINDING THE DOMAIN OF A RATIONAL FUNCTION

Recall that the domain of a function is simply a list of all x-values that you are allowed to substitute in the function. In a rational function, instead of trying to determine all the x-values that are included in the domain of the function, it is easier to determine which ones, if any, are not included in the domain of the function. Basically, you want to find the values of x that make the denominator equal to 0, and then the domain will be all real numbers except the x-values you just found. Let's look at some examples.

Example 3

Find the domain of $f(x) = \frac{5}{x}$.

Set the denominator equal to 0 and solve for x:

$x = 0$

Answer: The domain is all real numbers except 0.

Example 4

Find the domain of $f(x) = \frac{6}{2x - 3}$.

Set the denominator equal to 0 and solve for x:

$2x - 3 = 0$

$2x = 3$

$x = \dfrac{3}{2}$

Answer: The domain is all real numbers except $\dfrac{3}{2}$.

Example 5

Find the domain of $f(x) = \dfrac{2x+1}{x^2 - x - 2}$.

Set the denominator equal to 0 and solve for x:

$x^2 - x - 2 = 0$	Factor completely.
$(x-2)(x+1) = 0$	Set each factor equal to 0.
$x - 2 = 0$ or $x + 1 = 0$	Solve.
$x = 2$ or $x = -1$	

Answer: The domain is all real numbers except -1 and 2.

Example 6

Find the domain of $f(x) = \dfrac{x^2 - 5}{x^2 + 4}$.

Set the denominator equal to 0 and solve for x:

$x^2 + 4 = 0$

Wait! This expression doesn't factor. (Notice that it is a **sum** of two squares, not a **difference** of two squares.) In fact, this denominator will never equal 0 because x^2 will never equal -4.

Answer: In this case, the domain is all real numbers.

SIMPLIFYING RATIONAL EXPRESSIONS

To simplify a rational expression, complete the following steps:

Step 1: If possible, completely factor the numerator and denominator of the rational expression.

Step 2: Cancel the factors that are common to the numerator and denominator. Monomials can be cancelled with monomials, and binomials can be cancelled with binomials. **You cannot cancel a monomial with part of a binomial.**

Example 7

Simplify the rational expression $\dfrac{24x^3 y^6}{36xy^8}$.

$\dfrac{24x^3 y^6}{36xy^8}$	Divide the numbers by 12.

$$\frac{2x^{3-1}\,y^{6-8}}{3}$$

Use the rules of exponents.

Answer: $\dfrac{2x^2 y^{-2}}{3} = \dfrac{2x^2}{3y^2}$

Example 8

Simplify the rational expression: $\dfrac{x^2 - 5x - 24}{x^2 - 9x + 8}$.

Step 1: Since the numerator and denominator are not monomials, and they are not identical polynomials, nothing can be cancelled yet. We must first factor both the numerator and denominator to see if we can cancel anything.

$$\frac{(x-8)(x+3)}{(x-8)(x-1)}$$

Step 2: Cancel the like factors in the numerator and denominator.

$$\frac{\cancel{(x-8)}(x+3)}{\cancel{(x-8)}(x-1)} = \frac{(x+3)}{(x-1)}$$

Since the numerator and denominator are not identical binomials, no further simplifying is possible.

Answer: $\dfrac{(x+3)}{(x-1)}$

Example 9

Simplify the rational expression: $\dfrac{27 + 9x}{x^2 - 9}$

Step 1: Factor the numerator and denominator:

$$\frac{9(3+x)}{(x+3)(x-3)}$$

Step 2: Cancel like factors. Notice the numerator has a term $(3 + x)$ and the denominator has a term $(x + 3)$. Since the order of how we write addition does not matter (commutative property), these two binomials are actually equal and can be cancelled.

$$\frac{9\cancel{(3+x)}}{\cancel{(x+3)}(x-3)} = \frac{9}{x-3}$$

Since the numerator is a monomial, and the denominator is a binomial, no further simplification is possible.

Answer: $\dfrac{9}{x-3}$

Example 10

Simplify $\dfrac{x^2 - x}{2x^3 - 2x^4}$.

Step 1: Factor the numerator and denominator:

$$\dfrac{x(x-1)}{2x^3(1-x)}$$

Step 2: Cancel like factors. At first glance it appears that we can only cancel x into $2x^3$; however, the two binomials can also be cancelled with a little extra factoring. Notice $x - 1$ and $1 - x$ term are **the same binomials with opposite signs on each term**. If we rewrite them in the same order, we see the first term is $x - 1$, and the second term is $-x + 1$. What we need to do is factor an additional -1 from the numerator, and we will have the following expression.

$$\dfrac{-1x(-x+1)}{2x^3(1-x)}$$

The order of the binomial in the numerator can be rewritten to match the binomial in the dominator.

$$\dfrac{-1\cancel{x}(\cancel{1-x})}{2x^{\cancel{3}2}(\cancel{1-x})} = \dfrac{-1}{2x^2}$$

Answer: After canceling, we obtain the final answer of $\dfrac{-1}{2x^2}$.

RECOGNIZING −1 WHEN YOU SEE IT

Some expressions that equal -1 are $\dfrac{x-2}{2-x}, \dfrac{3-2x}{2x-3}, \dfrac{x-y}{y-x}$.

Some expressions that **don't** equal -1 are $\dfrac{x-2}{2+x}, \dfrac{3+2x}{2x-3}, \dfrac{x-y}{x+y}$.

From now on, when you see binomials in the numerator and denominator that are equal to -1, just cancel them, and put -1 in the numerator.

Example 11

Simplify: $\dfrac{3x^2 - 2x - 8}{12x + 16}$

Step 1: Factor.

$$\dfrac{(3x+4)(x-2)}{4(3x+4)}$$

Step 2: Cancel like terms.

$$\frac{(3x+4)(x-2)}{4(3x+4)} = \frac{(x-2)}{4}$$

Answer: No further reduction is possible; therefore, the final answer is $\frac{(x-2)}{4}$.

Practice Problems

1. Evaluate the following expression when $x = 2$: $\frac{x^2 - 4x}{2x + 1}$.

2. Evaluate the following expression when $x = -1$: $\frac{4 - x}{x^2 - 3x + 2}$.

Find the domain of each of the following rational functions:

3. $f(x) = \frac{5}{x}$

4. $f(x) = \frac{6}{5x^2}$

5. $f(x) = \frac{6}{x - 5}$

6. $f(x) = \frac{3x}{2x - 3}$

7. $f(x) = \frac{3x}{x^2 - 9}$

8. $f(x) = \frac{x - 5}{3x^2 - 3}$

9. $f(x) = \frac{x}{x^2 - x - 12}$

10. $f(x) = \frac{x + 2}{2x^2 + 5x + 3}$

Reduce the following to lowest terms.

11. $\frac{3x^2y^4}{12xy^3}$

12. $\frac{5xy}{20x^4y}$

13. $\dfrac{2x-4}{6}$

14. $\dfrac{x-3}{6-2x}$

15. $\dfrac{2x^2}{4x^3-10x}$

16. $\dfrac{5xy}{10x^2-15y^2}$

17. $\dfrac{x^2-9}{x^2+6x+9}$

18. $\dfrac{4x-x^2}{x^2-16}$

19. $\dfrac{x-8}{x^2-6x-16}$

20. $\dfrac{x^2-5x-6}{x^2-9x+18}$

21. $\dfrac{x^3-y^3}{x-y}$

22. $\dfrac{x^3-8}{x^2+2x+4}$

23. $\dfrac{2x^2+x-3}{9-4x^2}$

24. $\dfrac{2x^2-3x-20}{3x^2-11x-4}$

25. $\dfrac{x^3-4x^2+3x-12}{x-4}$

26. $\dfrac{x^3+3x^2+5x+15}{x^2y+2x^2+5y+10}$

Section 2
Multiplication and Division of Rational Expressions

Learning Objectives

When you finish your study of this section, you should be able to
- Multiply and divide rational expressions

Multiplication of Rational Expressions

The multiplication of rational expressions involves steps that are similar to multiplying fractions:

Step 1: Completely factor the numerator and denominator of each rational expression.

Step 2: Cancel out the factors which are common to the numerator and denominator of either fraction.

Step 3: Rewrite the final answer as a single fraction. You may leave it in factored form.

Why don't we start with an example that reviews how we multiply fractions, and then we will progress to more difficult rational expressions?

Example 1

Multiply: $\dfrac{2}{3} \cdot \dfrac{9}{5}$.

$\dfrac{2 \cdot 9}{3 \cdot 5}$ Multiply numerators. Multiply denominators.

$\dfrac{18}{15}$ Reduce to lowest terms.

Answer: $\dfrac{6}{5}$

A shorter way to do this problem is to "cancel" first:

$\dfrac{2}{\cancel{3}_1} \cdot \dfrac{\cancel{9}^{\,3}}{5}$ 3 goes into 9 three times. 3 goes into 3 once.

Answer: $\dfrac{6}{5}$

Now, let's multiply some rational expressions.

Example 2

Multiply: $\dfrac{4x^3}{5y} \cdot \dfrac{15}{14x}$.

$\dfrac{\cancel{4}2 x^{\cancel{3}2}}{\cancel{5}y} \cdot \dfrac{\cancel{15}3}{\cancel{14}7\cancel{x}}$ 5 goes into 15 three times. 2 goes into 4 twice. 2 goes into 14 seven times.

$x^3 \div x = x^2$

$\dfrac{2x^2}{y} \cdot \dfrac{3}{7}$ Multiply numerators together and denominators together.

Answer: $\dfrac{6x^2}{7y}$

Example 3

Multiply: $\dfrac{x^2 - 4}{x^2 - 25} \cdot \dfrac{x^2 - 3x - 10}{2x + 4}$.

Step 1: Factor the numerator and denominator of both fractions:

$\dfrac{(x+2)(x-2)}{(x+5)(x-5)} \cdot \dfrac{(x-5)(x+2)}{2(x+2)}$

Step 2: Cancel all like terms. A term in the numerator of either fraction can be cancelled with a term in the denominator of either fraction.

$\dfrac{(\cancel{x+2})(x-2)}{(x+5)(\cancel{x-5})} \cdot \dfrac{(\cancel{x-5})(x+2)}{2(\cancel{x+2})}$

$\dfrac{(x-2)}{(x+5)} \cdot \dfrac{(x+2)}{2}$

Step 3: Combine remaining terms into a single fraction in factored form.

Answer: $\dfrac{(x-2)(x+2)}{2(x+5)}$

Note: Don't bother multiplying out the terms on the top. Factored form is acceptable.

Example 4

Multiply: $\dfrac{2x^2 - 8x}{x^2 - 2x - 3} \cdot \dfrac{5x^2 + 4x - 1}{20x^2 - 84x + 16}$.

Step 1: Factor the numerator and denominator completely.

$$\dfrac{2x(x - 4)}{(x - 3)(x + 1)} \cdot \dfrac{(x + 1)(5x - 1)}{4(x - 4)(5x - 1)}$$

Step 2: Cancel like terms.

$$\dfrac{\cancel{2}x\cancel{(x - 4)}}{(x - 3)\cancel{(x + 1)}} \cdot \dfrac{\cancel{(x + 1)}\cancel{(5x - 1)}}{\cancel{4}2\cancel{(x - 4)}\cancel{(5x - 1)}}$$

$$\dfrac{x}{x - 3} \cdot \dfrac{1}{2}$$

Step 3: Combine remaining terms into a single fraction in factored form.

Answer: $\dfrac{x}{2(x - 3)}$

Division of Rational Expressions

If you remember the main difference between multiplying and dividing fractions, then you also know the difference between multiplying and dividing rational expressions. Basically, you are adding one step to the process, which is Step 1 below.

Step 1: Rewrite the division problem as a multiplication problem. Change the division symbol to multiplication, and write the reciprocal of the second expression (in other words, flip it over).

Step 2: Completely factor the numerator and denominator of each rational expression.

Step 3: Cancel out the factors which are common.

Step 4: Rewrite the final answer as a single fraction in factored form.

Again, let's look at a numerical example first.

Example 5

Divide: $\dfrac{5}{3} \div \dfrac{10}{7}$.

Step 1: Rewrite the problem without the division symbol by taking the reciprocal of the second fraction.

$$\dfrac{5}{3} \cdot \dfrac{7}{10}$$

Steps 2 & 3: Since there is not a polynomial to factor, just cancel denominators with numerators where appropriate.

$$\dfrac{\cancel{5}^{\,1}}{3} \cdot \dfrac{7}{\cancel{10}_{\,2}}$$

Step 4: Multiply the remaining terms and write the final answer as a single fraction.

Answer: $\dfrac{1}{3} \cdot \dfrac{7}{2} = \dfrac{7}{6}$

Example 6

Divide: $\dfrac{x-6}{3x+6y} \div \dfrac{x^2-4x-12}{x+2y}$

Step 1: Rewrite the division problem as a multiplication problem. Remember to take the reciprocal of the second fraction.

$$= \dfrac{x-6}{3x+6y} \cdot \dfrac{x+2y}{x^2-4x-12}$$

Step 2: Factor numerators and denominators completely.

$$= \dfrac{x-6}{3(x+2y)} \cdot \dfrac{x+2y}{(x-6)(x+2)}$$

Step 3: Cancel like terms.

$$= \dfrac{\cancel{x-6}\,^1}{3\cancel{(x+2y)}} \cdot \dfrac{\cancel{x+2y}\,^1}{\cancel{(x-6)}(x+2)} = \dfrac{1}{3} \cdot \dfrac{1}{(x+2)}$$

Step 4: Rewrite as a single fraction in factored form.

Answer: $\dfrac{1}{3(x+2)}$

Example 7

Divide: $\dfrac{x^2 + 7x + 12}{6x^5} \div \dfrac{x^2 - 16}{4x^5 - x^6}$.

Step 1: Rewrite the division problem as a multiplication problem.

$$= \dfrac{x^2 + 7x + 12}{6x^5} \cdot \dfrac{4x^5 - x^6}{x^2 - 16}$$

Step 2: Factor numerators and denominators completely.

$$= \dfrac{(x+3)(x+4)}{6x^5} \cdot \dfrac{x^5(4-x)}{(x+4)(x-4)}$$

Step 3: Cancel like terms. (Remember the rule for cancelling $x - 4$ and $4 - x$ from the previous section.)

$$= \dfrac{(x+3)\cancel{(x+4)}}{6\cancel{x^5}} \cdot \dfrac{\cancel{x^5}\cancel{(4-x)}^{-1}}{\cancel{(x+4)}\cancel{(x-4)}\,1} = \dfrac{(x+3)}{6} \cdot \dfrac{-1}{1}$$

Step 4: Rewrite as a single fraction in factored form.

Answer: $\dfrac{-(x+3)}{6}$

Practice Problems

Perform the indicated operations and reduce to lowest terms.

1. $\dfrac{6}{5} \cdot \dfrac{1}{8}$

2. $\dfrac{5}{6} \cdot \dfrac{12}{5}$

3. $\dfrac{3x^2}{7} \cdot \dfrac{10}{9x}$

4. $\dfrac{5xy^3}{4} \cdot \dfrac{10x}{y^2}$

5. $\dfrac{x+2}{6x} \cdot \dfrac{x^5}{x^2-4}$

6. $\dfrac{y}{y-5} \cdot \dfrac{25-y^2}{2y}$

7. $\dfrac{4x+8}{5} \cdot \dfrac{5x+15}{x^2+5x+6}$

8. $\dfrac{x^2-x-6}{x+4} \cdot \dfrac{x^2-16}{x-3}$

9. $\dfrac{x-2}{x^2-x} \cdot \dfrac{x^2-1}{x^2-4x+4}$

10. $\dfrac{x^2-4x-5}{x^2-25} \cdot \dfrac{x^2+7x+10}{5-x}$

11. $\dfrac{2x^2+9x+4}{4x^2-1} \cdot \dfrac{10x-5}{x^2+9x+20}$

12. $\dfrac{3x^2+x-2}{15x^2-6x} \cdot \dfrac{5x^2-7x+2}{x^2-1}$

Perform the indicated operations and reduce to lowest terms.

13. $\dfrac{2}{3} \div \dfrac{4}{9}$

14. $\dfrac{5x}{8y} \div \dfrac{x^2}{4y}$

15. $\dfrac{x-2}{y^3} \div \dfrac{5y}{2x-4}$

16. $\dfrac{5-x}{6} \div \dfrac{x^2-25}{10}$

17. $\dfrac{2x+6}{4x} \div \dfrac{x^2-9}{2(x-3)}$

18. $\dfrac{x^2+7x+12}{6x} \div \dfrac{x^2-9}{12x}$

19. $\dfrac{3x^2-6x^3}{x^2-9y^2} \div \dfrac{6x^3}{x+3y}$

20. $\dfrac{5x^2 + 7x - 6}{x^2 - 4x + 3} \div \dfrac{3x^2 + 5x - 2}{4x^2 - 12x}$

21. $\dfrac{3x}{5} \cdot \dfrac{x^2}{15} \div \dfrac{x^4}{25}$

22. $\dfrac{2y}{7} \div \dfrac{4}{21y^2} \cdot \dfrac{1}{9y}$

Section 3
Adding and Subtracting Rational Expressions

Learning Objectives

When you finish your study of this section, you should be able to
- Add and subtract rational expressions with like denominators
- Find least common denominators (LCDs)
- Add and subtract rational expressions with unlike denominators

Addition and Subtraction of Rational Expressions With Like Denominators

Adding and subtracting rational expressions is very similar to adding and subtracting fractions. Basically, you combine the numerators, keep the denominator, and then simplify if possible. The formal steps follow:

Step 1: Add or subtract the numerators. Place the result over the common denominator.

Step 2: Simplify the rational expression if possible; that is, reduce your answer to lowest terms.

Example 1

Add: $\dfrac{5x}{6} + \dfrac{7}{6}$.

Step 1: Add the numerators. Since the numerators are not like terms, you obtain the following.

$$\dfrac{5x + 7}{6}$$

Step 2: Simplify. Since there is nothing that can be simplified, the final answer is the following.

Answer: $\dfrac{5x + 7}{6}$

Example 2

Subtract: $\dfrac{2x^2 - 6x + 9}{x^2 - 9} - \dfrac{x^2 - 7x + 15}{x^2 - 9}$.

Step 1: Subtract the numerators by combining like terms.

$$\dfrac{(2x^2 - 6x + 9) - (x^2 - 7x + 15)}{x^2 - 9}$$

Keep in mind, when subtracting polynomials, you must change the sign of **each term** in the second polynomial.

$\dfrac{2x^2 - 6x + 9 - x^2 + 7x - 15}{x^2 - 9}$ Distribute the negative sign.

$\dfrac{x^2 + x - 6}{x^2 - 9}$ Combine like terms.

Step 2: Simplify the rational expression. Since both the numerator and denominator can be factored, we need to factor to see if something will cancel.

$\dfrac{(x + 3)(x - 2)}{(x + 3)(x - 3)}$ This is the factored form.

$\dfrac{\cancel{(x + 3)}(x - 2)}{\cancel{(x + 3)}(x - 3)}$ Cancel like terms.

Answer: $\dfrac{x - 2}{x - 3}$

Example 3

Add: $\dfrac{x + 1}{2x^2 - 5x - 12} + \dfrac{x + 2}{2x^2 - 5x - 12}$.

Step 1: Add the numerators by combining like terms.

$$\dfrac{(x + 1) + (x + 2)}{2x^2 - 5x - 12} = \dfrac{2x + 3}{2x^2 - 5x - 12}$$

Step 2: Simplify if possible. Notice the denominator can be factored.

$$\frac{2x+3}{(2x+3)(x-4)}$$ 	This is the factored form.

$$\frac{\cancel{2x+3}}{\cancel{(2x+3)}(x-4)}$$ 	Cancel like terms.

Answer: $\dfrac{1}{x-4}$

Finding the Least Common Denominator (LCD)

The least common denominator (LCD) of two or more rational expressions is a polynomial of least degree whose factors include all the factors of the denominators in the group. To find the LCD, use the following steps.

Step 1: Factor each denominator.

If the denominators are monomials,
- List the multiples for each number.
- The least common multiple (LCM) is the smallest number that appears in both lists.

If the denominators are polynomials, factor each denominator completely (if possible).

Step 2: Select every unique factor. If a factor appears in more than one denominator, select the factor with the largest exponent.

Step 3: Multiply the factors. (If the factors are polynomials, we do not need to multiply them.)

Example 4

Find the LCD of the following: $\dfrac{5}{6xy}$, $\dfrac{1}{18x^2y}$.

Step 1: Since the denominators are monomials, find the least common multiple (LCM) of the numbers 6 and 18:

Multiples of 6: 		6, 12, $\underline{18}$, 24, 30, 36...

Multiples of 18: 		$\underline{18}$, 36, 54, 72, 90...

The LCM is 18.

Hint: If the smaller number divides evenly into the larger number, you can always use that larger number as the LCM. Since 6 divides evenly into 18 three times, that would tell us to use 18 as the LCM.

Step 2: Keep the **largest** exponent of each variable.

The largest exponent of x is 2.

The largest exponent of y is 1.

Step 3: Multiply 18, x^2, and y together to obtain the LCD.

Answer: Thus, the LCD is $18x^2y$.

Note: In other words, the LCD consists of the LCM of all of the numbers in the denominators, multiplied by the largest exponent of each variable.

Example 5

Find the LCD: $\frac{4}{x}, \frac{3x}{x+5}$.

Answer: Since one factor is a binomial, and the other is a monomial, the LCD is the product of both: $x(x+5)$.

Note: We don't actually multiply the terms together since we will leave our final answers in factored form anyway.

Example 6

Find the LCD : $\frac{x+3}{x^2+2x-15}, \frac{x}{x^2+6x+5}$.

Step 1: Factor each denominator.

$x^2 + 2x - 15 = (x+5)(x-3)$

$x^2 + 6x + 5 = (x+1)(x+5)$

Step 2: Select every unique factor appearing. Since there are two $(x+5)$ terms with the same power of 1, we take just one of them.

$(x+1), (x-3), (x+5)$

Step 3: Write the factors as a product, but do not multiply them.

Answer: $(x+1)(x-3)(x+5)$

Example 7

Find the LCD: $\frac{5}{2x+4}, \frac{3}{x^2-4}$.

Step 1: Factor each denominator.

$2x + 4 = 2(x+2)$

$x^2 - 4 = (x+2)(x-2)$

Step 2: Select every unique factor appearing.

$2, (x + 2), (x - 2)$

Step 3: Write the factors as a product, but do not multiply them.

Answer: $2(x + 2)(x - 2)$

Addition and Subtraction of Rational Expressions With Unlike Denominators

The addition and subtraction of rational expressions with unlike denominators is again similar to adding and subtracting fractions with unlike denominators. Just follow these steps:

Step 1: Find the LCD of the rational expressions.

Step 2: Write each rational expression as an equivalent rational expression with the LCD as its denominator.

Step 3: Add or subtract the numerators. Place the result over the common denominator.

Step 4: Reduce the rational expression to lowest terms, if possible.

Try some examples:

Example 8
Add the rational expressions: $\dfrac{4x}{5} + \dfrac{3}{2x}$.

Step 1: Find the LCD of the rational expressions.

The LCM of 5 and 2 is 10.

The largest exponent of x is 1, so the LCD is $10x$.

Step 2: Rewrite each as an equivalent expression.

$$\frac{4x}{5} = \frac{4x(2x)}{5(2x)} = \frac{8x^2}{10x}$$

$$\frac{3}{2x} = \frac{3(5)}{2x(5)} = \frac{15}{10x}$$

Step 3: Add the numerators. Since the numerators are not like terms, we get the following:

$$\frac{8x^2}{10x} + \frac{15}{10x} = \frac{8x^2 + 15}{10x}$$

Step 4: Simplify, if possible. This expression cannot be simplified.

Answer: $\dfrac{8x^2 + 15}{10x}$

Example 9

Add the rational expressions: $\dfrac{3}{x} + \dfrac{5}{x+4}$

Step 1: Find the LCD of the rational expressions.

The LCD of x and $x + 4$ is $x(x + 4)$.

Step 2: Rewrite each as an equivalent expression.

$$\dfrac{3}{x} = \dfrac{3(x+4)}{x(x+4)} = \dfrac{3x+12}{x(x+4)}$$

$$\dfrac{5}{x+4} = \dfrac{5(x)}{x(x+4)} = \dfrac{5x}{x(x+4)}$$

The problem rewritten becomes

$$\dfrac{3x+12}{x(x+4)} + \dfrac{5x}{x(x+4)}$$

Step 3: Add the numerators.

$$\dfrac{8x+12}{x(x+4)}$$

Step 4: Even though the numerator can be factored into $4(2x + 3)$, nothing can be cancelled within the denominator.

Answer: $\dfrac{8x+12}{x(x+4)}$ or $\dfrac{4(2x+3)}{x(x+4)}$

Example 10

Subtract: $\dfrac{2}{x^2 - 4} - \dfrac{3}{x^2 + 4x + 4}$

Step 1: Find the LCD of the rational expressions by factoring each denominator first.

$x^2 - 4 = (x-2)(x+2)$

$x^2 + 4x + 4 = (x+2)(x+2) = (x+2)^2$

Thus, the LCD is $(x-2)(x+2)^2$. (Remember that you keep the larger exponent.)

Step 2: Write each equivalent fraction.

$$\frac{2}{(x-2)(x+2)} = \frac{2(x+2)}{(x-2)(x+2)^2} = \frac{2x+4}{(x-2)(x+2)^2}$$

$$\frac{3}{(x+2)^2} = \frac{3(x-2)}{(x-2)(x+2)^2} = \frac{3x-6}{(x-2)(x+2)^2}$$

The problem rewritten becomes

$$\frac{2x+4}{(x-2)(x+2)^2} - \frac{3x-6}{(x-2)(x+2)^2}$$

Step 3: Subtract the numerators. (Remember to distribute the negative sign.) Place the result over the common denominator.

Answer: $\frac{2x+4-3x+6}{(x-2)(x+2)^2} = \frac{-x+10}{(x-2)(x+2)^2}$

Practice Problems

Perform the indicated operation and simplify.

1. $\frac{2}{3} + \frac{2}{3}$

2. $\frac{6}{5} - \frac{11}{5}$

3. $\frac{6x}{7} + \frac{8x}{7}$

4. $\frac{2y}{7} - \frac{5y}{7}$

5. $\frac{5}{8} - \frac{x-3}{8}$

6. $\frac{3x^2-x}{4} + \frac{3x+6}{4}$

7. $\dfrac{x^2 - x}{x - 1} - \dfrac{x - 1}{x - 1}$

8. $\dfrac{3x^2 + 4x - 3}{x + 2} - \dfrac{x^2 + x - 1}{x + 2}$

9. $\dfrac{1}{2} - \dfrac{2}{3}$

10. $\dfrac{5}{6} + \dfrac{4}{5}$

11. $\dfrac{x}{3} + \dfrac{6}{x}$

12. $\dfrac{y}{5} - \dfrac{4}{3y}$

13. $\dfrac{4x}{5y} + \dfrac{9y}{4x^2}$

14. $\dfrac{x}{2y} - \dfrac{4y}{3x^3}$

15. $\dfrac{8}{5y} + 3$

16. $5 - \dfrac{4}{x}$

17. $\dfrac{5}{x} - \dfrac{2}{x - 4}$

18. $\dfrac{x}{3} + \dfrac{6}{x - 3}$

19. $\dfrac{1}{x - 2} + \dfrac{3}{x - 1}$

20. $\dfrac{x}{x + 5} - \dfrac{4}{x + 2}$

21. $\dfrac{x - 5}{3} + \dfrac{2}{x - 7}$

22. $\dfrac{x - 1}{x + 2} - \dfrac{4}{2x}$

23. $\dfrac{2}{x-3} + \dfrac{5}{3-x}$

24. $\dfrac{x+4}{x-4} - \dfrac{5}{4-x}$

25. $\dfrac{x+2}{x-3} - \dfrac{x-1}{x-2}$

26. $\dfrac{x}{x+4} + \dfrac{2x-1}{x+1}$

27. $\dfrac{4}{x^2-9} - \dfrac{2}{x+3}$

28. $\dfrac{1}{x+2} + \dfrac{x}{x^2-4}$

29. $\dfrac{x-5}{2x-4} + \dfrac{3x}{4}$

30. $\dfrac{5}{3x^2-3x} - \dfrac{2x}{x^2-1}$

31. $\dfrac{5}{x^2-1} + \dfrac{x}{x^2+4x-5}$

32. $\dfrac{4}{x^2-3x+2} - \dfrac{2}{x^2-8x+12}$

33. $\dfrac{2}{x} - \dfrac{5}{x-1} + \dfrac{3}{x+1}$

34. $\dfrac{9}{5} - 2x + \dfrac{6}{2x-3}$

Section 4
Complex Fractions

Learning Objectives

When you finish your study of this section, you should be able to
- Recognize complex fractions
- Simplify complex fractions

Complex Fractions

A complex fraction is a fraction in which the numerator and/or denominator consist(s) of a rational expression. Complex fractions should always be simplified; do not leave them in their original form.
Examples of complex fractions:

a. $\dfrac{\dfrac{1}{y^2} - 1}{1 + \dfrac{1}{y}}$
b. $\dfrac{\dfrac{x}{8}}{\dfrac{3x}{10}}$
c. $\dfrac{2 - \dfrac{1}{3}}{8 + \dfrac{2}{3}}$
d. $\dfrac{\dfrac{8}{4}}{x - 3}$

Simplifying Complex Fractions by Dividing

Whenever you have a complex fraction that contains only division (such as examples b and d above), then use the following steps to simplify the expression:

Step 1: Rewrite the complex fraction as a division of two fractions (horizontally).

Step 2: Then, follow the same steps as outlined for the division of rational expressions (see Section 2 of this chapter).

Example 1

Simplify: $\dfrac{\dfrac{x}{8}}{\dfrac{3x}{10}}$.

Step 1: Rewrite this fraction as a division problem.

$$\frac{x}{8} \div \frac{3x}{10}$$

Step 2: Now, follow the steps for the division of rational expressions.

$\frac{x}{8} \cdot \frac{10}{3x}$ Rewrite as a multiplication problem.

$\frac{\cancel{x}1}{\cancel{8}4} \cdot \frac{\cancel{10}5}{3\cancel{x}}$ x goes into x once. 2 goes into 8 four times. 2 goes into 10 five times.

$\frac{1}{4} \cdot \frac{5}{3}$

Answer: $\frac{5}{12}$

Example 2

Simplify: $\dfrac{\dfrac{5}{x^2 - 9}}{\dfrac{15}{3x + 9}}.$

$\dfrac{5}{x^2 - 9} \div \dfrac{15}{3x + 9}$

$\dfrac{5}{x^2 - 9} \cdot \dfrac{3x + 9}{15}$ Rewrite as a multiplication problem.

$\dfrac{\cancel{5}1}{(\cancel{x+3})(x-3)} \cdot \dfrac{\cancel{3}(\cancel{x+3})1}{\cancel{15}1}$

$\dfrac{1}{x - 3} \cdot \dfrac{1}{1}$ Multiply numerators and denominators.

Answer: $\dfrac{1}{x - 3}$

Simplifying Complex Fractions by Using a LCD

To simplify more complicated complex fractions, use the following steps:
Step 1: Multiply the numerator and denominator by the LCD of all the denominators in the original complex fraction.

Step 2: Then, reduce the remaining fraction.

That's all there is to it.

Example 3

Simplify: $\dfrac{2 - \dfrac{1}{3}}{8 + \dfrac{2}{3}}$.

The denominators in the top and the bottom are both 3, so that is your LCD. Multiply the top and bottom by 3.

$\dfrac{3\left(2 - \dfrac{1}{3}\right)}{3\left(8 + \dfrac{2}{3}\right)}$ Make sure to multiply each term by 3.

$\dfrac{3(2) - 3\left(\dfrac{1}{3}\right)}{3(8) + 3\left(\dfrac{2}{3}\right)}$ Multiply.

$\dfrac{6 - 1}{24 + 2}$ Simplify.

Answer: $\dfrac{5}{26}$

Example 4

Simplify: $\dfrac{\dfrac{1}{y^2} - 1}{1 + \dfrac{1}{y}}$.

In this problem, the denominator in the top is y^2, and the denominator in the bottom is y, so the LCD is y^2. Multiply top and bottom by y^2.

$\dfrac{y^2\left(\dfrac{1}{y^2} - 1\right)}{y^2\left(1 + \dfrac{1}{y}\right)} = \dfrac{y^2\left(\dfrac{1}{y^2}\right) - y^2(1)}{y^2(1) + y^2\left(\dfrac{1}{y}\right)}$ Multiply each term by y^2.

$\dfrac{1 - y^2}{y^2 + y}$ Can this fraction be simplified? Yes. Both the numerator and denominator can be factored.

$$\frac{(1+y)(1-y)}{y(y+1)}$$ This is the factored form.

$$\frac{\cancel{(1+y)}(1-y)}{y\cancel{(y+1)}}$$ Cancel the common factor.

Answer: $\dfrac{1-y}{y}$

Example 5

Simplify: $\dfrac{2 - \dfrac{3}{y+1}}{1 + \dfrac{2}{y-1}}$.

This time the LCD is $(y+1)(y-1)$. Multiply top and bottom by the LCD.

$$\frac{(y+1)(y-1)\left(2 - \dfrac{3}{y+1}\right)}{(y+1)(y-1)\left(1 + \dfrac{2}{y-1}\right)}$$ Multiply each term by the LCD.

$$\frac{(y+1)(y-1)(2) - \cancel{(y+1)}(y-1)\left(\dfrac{3}{\cancel{y+1}}\right)}{(y+1)(y-1)(1) + (y+1)\cancel{(y-1)}\left(\dfrac{2}{\cancel{y-1}}\right)}$$ Cancel common terms. Multiply.

$$\frac{(y^2-1)(2) - (y-1)(3)}{(y^2-1)(1) + (y+1)(2)}$$ Multiply some more.

$$\frac{2y^2 - 2 - 3y + 3}{y^2 - 1 + 2y + 2}$$ Combine like terms.

$$\frac{2y^2 - 3y + 1}{y^2 + 2y + 1}$$ Factor.

Answer: $\dfrac{(2y-1)(y-1)}{(y+1)(y+1)}$ Nothing cancels.

Chapter 6: Rational Expressions and Equations

Practice Problems

Simplify the following.

1. $\dfrac{\frac{3}{5}}{\frac{1}{10}}$

2. $\dfrac{\frac{5}{6}}{\frac{7}{12}}$

3. $\dfrac{\frac{x}{x-2}}{\frac{5}{4x-8}}$

4. $\dfrac{\frac{2x}{x+5}}{\frac{6}{x-5}}$

5. $\dfrac{\frac{x-1}{x+1}}{\frac{x^2-1}{1-x^2}}$

6. $\dfrac{\frac{5y}{2x+1}}{\frac{10y^2}{2x^2+x-1}}$

7. $\dfrac{1-\frac{2}{3}}{4-\frac{5}{6}}$

8. $\dfrac{3+\frac{2}{5}}{4-\frac{3}{10}}$

9. $\dfrac{\frac{1}{5}+\frac{2}{3}}{\frac{5}{3}-\frac{1}{15}}$

10. $\dfrac{\frac{1}{2}-\frac{1}{8}}{\frac{7}{8}+\frac{3}{4}}$

11. $\dfrac{\dfrac{x}{2}+\dfrac{3}{5}}{\dfrac{5}{2}-\dfrac{x}{5}}$

12. $\dfrac{\dfrac{y}{3}+\dfrac{5}{6}}{\dfrac{y}{2}-\dfrac{1}{12}}$

13. $\dfrac{\dfrac{1}{x^2}}{\dfrac{3}{x}-\dfrac{5}{4}}$

14. $\dfrac{\dfrac{1}{y}+\dfrac{1}{x}}{\dfrac{3}{xy}}$

15. $\dfrac{\dfrac{1}{x}+\dfrac{1}{y}}{\dfrac{x}{y}-\dfrac{y}{x}}$

16. $\dfrac{\dfrac{5}{2x}-\dfrac{4}{3}}{\dfrac{1}{12}+\dfrac{3}{x^2}}$

17. $\dfrac{\dfrac{x-1}{4}}{\dfrac{1}{2}+\dfrac{3}{x-2}}$

18. $\dfrac{1-\dfrac{4}{x^2}}{\dfrac{x}{x-2}}$

19. $\dfrac{3+\dfrac{5}{x+3}}{5+\dfrac{3}{x+5}}$

20. $\dfrac{4-\dfrac{1}{x+2}}{2-\dfrac{3}{x-2}}$

21. $\dfrac{\dfrac{5}{y}-\dfrac{2}{y+1}}{\dfrac{3}{y+1}+\dfrac{4}{y^2}}$

22. $\dfrac{\dfrac{3}{x} - \dfrac{2}{x-1}}{\dfrac{4}{x} + \dfrac{1}{x+1}}$

SECTION 5
DIVISION OF POLYNOMIALS

Learning Objectives

When you finish your study of this section, you should be able to
- Divide a polynomial by a monomial
- Perform polynomial long division
- Perform synthetic division

Dividing a Polynomial by a Monomial

To divide a polynomial by a monomial, divide each term of the polynomial by the monomial. Let's look at some examples:

Example 1

Divide: $(6a^2 + 4a - 8) \div 2a$.

Rewrite as a fraction: $\dfrac{6a^2 + 4a - 8}{2a}$

$\dfrac{6a^2}{2a} + \dfrac{4a}{2a} - \dfrac{8}{2a}$ Divide each term by the monomial.

$3a^{2-1} + 2a^{1-1} - \dfrac{4}{a}$ Use the quotient rule for exponents.

$3a + 2a^0 - \dfrac{4}{a}$ $a^0 = 1$.

$3a + 2(1) - \dfrac{4}{a}$

Answer: $3a + 2 - \dfrac{4}{a}$

Chapter 6: Rational Expressions and Equations

Example 2

Divide: $\dfrac{4xy^2 - 10x^2y + 6xy}{2xy}$.

$\dfrac{4xy^2}{2xy} - \dfrac{10x^2y}{2xy} + \dfrac{6xy}{2xy}$ Divide each term by the monomial.

$2y^{2-1} - 5x^{2-1} + 3$ Use the quotient rule for exponents.

Answer: $2y - 5x + 3$

Long Division of Polynomials

When we divide a polynomial by a polynomial, we need to use a process that is similar to long division of numbers.

Step 1: Rewrite the problem using long division. Write the inside and outside polynomial in decreasing order of powers.

Step 2: Divide the first term of the inside polynomial by the first term of the outside polynomial. Place answer above long division symbol.

Step 3: Multiply the answer from Step 2 by the outside polynomial. Place this answer below the inside polynomial.

Step 4: Subtract the inside polynomials. Change the sign of each term in the bottom row and combine like terms.

Step 5: Repeat Steps 2 through 4 as necessary.

Example 3

Find the quotient: $(x^2 - x - 12) \div (x - 4)$.

Step 1: Rewrite the problem, using the long division form. Notice the polynomial $x - 4$ is written on the outside of the long division symbol and the polynomial $x^2 - x - 12$ is written on the inside.

$x - 4 \overline{)x^2 - x - 12}$

Step 2: Divide the first term of the inside polynomial by the first term of the outside polynomial. The first term on the inside, x^2, divided by the first term on the outside, x, results in the answer, x. Notice the answer, x, is placed above the long division symbol.

$\,x$
$x - 4 \overline{)x^2 - x - 12}$

Step 3: Multiply the answer from Step 2, which is x, by the outside polynomial, $x - 4$, resulting in the expression $x^2 - 4x$. This answer, $x^2 - 4x$, is placed below the inside polynomial.

$$\begin{array}{r} x \\ x-4\overline{\smash{\big)}\,x^2 - x - 12} \\ x^2 - 4x \end{array}$$

Step 4: Subtract the inside polynomials. Keep in mind that you must **change the sign** of each term in the second polynomial before combining like terms. Thus, the polynomial becomes $-x^2 + 4x$. Now, our like terms, positive x^2 and wnegative x^2, add to zero. Negative x added to $4x$ becomes $3x$. Next, bring down the -12.

$$\begin{array}{r} x \\ x-4\overline{\smash{\big)}\,x^2 - x - 12} \\ \underline{-x^2 + 4x} \\ 3x - 12 \end{array}$$

The final step is to repeat Steps 2 through 4 again. The first term on the inside, $3x$, divided by the first term on the outside, x, is 3. Now, multiply 3 by the outside polynomial, $x - 4$, resulting in the expression $3x - 12$. Subtracting the inside polynomials will result in a remainder of 0.

$$\begin{array}{r} x + 3 \\ x-4\overline{\smash{\big)}\,x^2 - x - 12} \\ \underline{-x^2 + 4x} \\ 3x - 12 \\ \underline{3x - 12} \\ 0 \end{array}$$

Answer: $x + 3$

Example 4

Find the quotient: $(3a^2 - 7a - 4) \div (a - 4)$.

Step 1: Rewrite the problem in long division form.

$$a - 4\overline{\smash{\big)}\,3a^2 - 7a - 4}$$

Step 2: Divide the first term of the inside polynomial by the first term of the outside polynomial. Place your answer above the long division symbol.

$$\begin{array}{r} 3a \\ a-4\overline{\smash{\big)}\,3a^2 - 7a - 4} \end{array}$$

Step 3: Multiply the answer for Step 2 by the outside polynomial. Place your answer below the inside polynomial.

$$a - 4 \overline{\smash{)}\begin{array}{r}3a\\ 3a^2 - 7a - 4\end{array}}$$
$$\underline{3a^2 - 12a}$$

$3a(a - 4) = 3a^2 - 12a$

Step 4: Subtract the inside polynomials. **Change the sign** of each term in the bottom row and combine like terms. Then, bring down the -4.

$$a - 4 \overline{\smash{)}\begin{array}{r}3a\\ 3a^2 - 7a - 4\end{array}}$$
$$\underline{-3a^2 - 12a}$$
$$5a - 4$$

Step 5: Repeat Steps 2 through 4.

$$a - 4 \overline{\smash{)}\begin{array}{r}3a + 5\\ 3a^2 - 7a - 4\end{array}}$$
$$\underline{-3a^2 + 12a}$$
$$5a - 4$$
$$5a - 20$$

$5(a - 4) = 5a - 20$

Change the signs and combine like terms.

$$a - 4 \overline{\smash{)}\begin{array}{r}3a + 5\\ 3a^2 - 7a - 4\end{array}}$$
$$\underline{-3a^2 + 12a}$$
$$5a - 4$$
$$\underline{-5a + 20}$$
$$16$$

The remainder is 16, which is written as a fraction.

Answer: $3a + 5 + \dfrac{16}{a - 4}$

Example 5

Find the quotient: $(x^3 - 4x + 3) \div (x + 1)$.

In this problem we are missing a power of x, the second power, so we add a zero term and write $x^3 - 4x + 3$ as $x^3 + 0x^2 - 4x + 3$. Adding the zero term will correctly align the polynomial terms for long division.

$$x + 1 \overline{\smash{)}x^3 + 0x^2 - 4x + 3}$$

$$x+1\overline{)x^3+0x^2-4x+3}^{\,x^2}$$
$$\phantom{x+1\overline{)}}\underline{x^3+x^2}$$

$x^2(x+1) = x^3+x^2$

$$x+1\overline{)x^3+0x^2-4x+3}^{\,x^2}$$
$$\phantom{x+1\overline{)}}\underline{-x^3-x^2}$$
$$\phantom{x+1\overline{)x^3+}}-x^2-4x$$

Change the signs and combine like terms.
Bring down the $-4x$.

$$x+1\overline{)x^3+0x^2-4x+3}^{\,x^2-x}$$
$$\phantom{x+1\overline{)}}\underline{-x^3-x^2}$$
$$\phantom{x+1\overline{)x^3+}}-x^2-4x$$
$$\phantom{x+1\overline{)x^3+}}-x^2-x$$

$-x(x+1) = -x^2-x$

$$x+1\overline{)x^3+0x^2-4x+3}^{\,x^2-x}$$
$$\phantom{x+1\overline{)}}\underline{-x^3-x^2}$$
$$\phantom{x+1\overline{)x^3+}}-x^2-4x$$
$$\phantom{x+1\overline{)x^3+}}\underline{+x^2+x}$$
$$\phantom{x+1\overline{)x^3+0x^2}}-3x+3$$

Change signs. Combine like terms. Bring down the $+3$.

$$x+1\overline{)x^3+0x^2-4x+3}^{\,x^2-x-3}$$
$$\phantom{x+1\overline{)}}\underline{-x^3-x^2}$$
$$\phantom{x+1\overline{)x^3+}}-x^2-4x$$
$$\phantom{x+1\overline{)x^3+}}\underline{+x^2+x}$$
$$\phantom{x+1\overline{)x^3+0x^2}}-3x+3$$
$$\phantom{x+1\overline{)x^3+0x^2}}-3x-3$$

$-3(x+1) = -3x-3$

$$x+1\overline{)x^3+0x^2-4x+3}^{\,x^2-x-3}$$
$$\phantom{x+1\overline{)}}\underline{-x^3-x^2}$$
$$\phantom{x+1\overline{)x^3+}}-x^2-4x$$
$$\phantom{x+1\overline{)x^3+}}\underline{+x^2+x}$$
$$\phantom{x+1\overline{)x^3+0x^2}}-3x+3$$
$$\phantom{x+1\overline{)x^3+0x^2}}\underline{+3x+3}$$
$$\phantom{x+1\overline{)x^3+0x^2-4x+}}6$$

Change signs. Combine like terms.

Answer: $x^2 - x - 3 + \dfrac{6}{x+1}$

Example 6

Find the quotient: $(3x^3 - 2x^2 + 5x - 2) \div (x^2 - x + 3)$.

$x^2 - x + 3 \overline{)3x^3 - 2x^2 + 5x - 2}$ Write as long division problem.

$$\begin{array}{r} 3x \\ x^2 - x + 3 \overline{)3x^3 - 2x^2 + 5x - 2} \\ 3x^3 - 3x^2 + 9x \end{array}$$

$3x(x^2 - x - 3) = 3x^3 - 3x^2 + 9x$

$$\begin{array}{r} 3x \\ x^2 - x + 3 \overline{)3x^3 - 2x^2 + 5x - 2} \\ -3x^3 + 3x^2 - 9x \\ x^2 - 4x - 2 \end{array}$$

Change signs. Combine like terms. Bring down the -2

$$\begin{array}{r} 3x + 1 \\ x^2 - x + 3 \overline{)3x^3 - 2x^2 + 5x - 2} \\ -3x^3 + 3x^2 - 9x \\ x^2 - 4x - 2 \\ x^2 - x + 3 \end{array}$$

$1(x^2 - x + 3) = x^2 - x + 3$

$$\begin{array}{r} 3x + 1 \\ x^2 - x + 3 \overline{)3x^3 - 2x^2 + 5x - 2} \\ -3x^3 + 3x^2 - 9x \\ x^2 - 4x - 2 \\ -x^2 + x - 3 \\ -3x - 5 \end{array}$$

Change signs. Combine like terms. This is your remainder.

Answer: $3x + 1 + \dfrac{-3x - 5}{x^2 - x + 3}$

Synthetic Division

Some long division problems can be solved in a shorter, more compact form called synthetic division. Division problems that involve a polynomial divided by a binomial in the form of $x - k$ can be solved by using synthetic division. Examples 3, 4, and 5 in this section could have been solved using this method. We are going to use synthetic division on Example 5 so that you can compare the two strategies on the same problem. Before we begin, consider the steps to use:

Step 1: If the divisor is in the form of $x - k$, then find k and write it on the outside of the division symbol.

Step 2: Write down the coefficients of the problem, without the variables, under the division symbol.

Step 3: Follow the procedure outlined in Example 7.

Let's work through an example carefully:

Example 7

Find the quotient: $(x^3 - 4x + 3) \div (x + 1)$. (This is Example 5 redone.)

We think of $(x + 1)$ as $(x - (-1))$, so in this problem, $k = -1$. That is placed on the outside.

$$-1 \;|\; 1 \quad 0 \quad -4 \quad 3$$

This row represents $1x^3 + 0x^2 - 4x + 3$.

$$-1 \;|\; 1 \quad 0 \quad -4 \quad 3$$
$$\underline{}$$
$$1$$

Bring down the 1.

$$-1 \;|\; 1 \quad 0 \quad -4 \quad 3$$
$$-1. \;\underline{-1}$$
$$1 \quad -1$$

Multiply 1 by the -1 on the outside. Add 0 to obtain

$$-1 \;|\; 1 \quad 0 \quad -4 \quad 3$$
$$\underline{-1 \quad 1}$$
$$1 \quad -1 \quad -3$$

$(-1)(-1) = 1$. Write 1 under -4. Then, $-4 + 1 = -3$

$$\begin{array}{r|rrrr} -1 & 1 & 0 & -4 & 3 \\ & & -1 & 1 & 3 \\ \hline & 1 & -1 & -3 & 6 \end{array}$$

$(-3)(-1) = 3$. Write 3 under 3. Then, $3 + 3 = 6$.

WRITING YOUR FINAL ANSWER

The degree of your quotient is always **one less** than the degree of the dividend; in this example, the dividend had 3 as its largest exponent, so the quotient will have 2 (**one less**) as its largest exponent. Thus, the 1 will represent the x^2 term, -1 will be the x term, -3 the constant term, and 6 is the remainder term. These numbers are the coefficients of the quotient.

Answer: $1x^2 - 1x - 3 + \dfrac{6}{x+1}$ or $x^2 - x - 3 + \dfrac{6}{x+1}$

Note: Even though we broke this example into a dozen lines of text, it could have been completed using three lines on your paper. Again, synthetic division can be used only when dividing by a binomial of the form $x - k$.

Example 8

Find the quotient: $(x^3 - 8) \div (x - 2)$.

$x^3 - 8 = 1x^3 + 0x^2 + 0x - 8$. We are dividing by $x - 2$, so $k = 2$.

$$\begin{array}{r|rrrr} 2 & 1 & 0 & 0 & -8 \\ & & 2 & & \\ \hline & 1 & & & \end{array}$$

Bring down the 1. Multiply 2 by 1. Write 2 under the first 0.

$$\begin{array}{r|rrrr} 2 & 1 & 0 & 0 & -8 \\ & & 2 & 4 & \\ \hline & 1 & 2 & & \end{array}$$

Add 0 and 2. Multiply 2 by 2. Write 4 under the second 0.

$$\begin{array}{r|rrrr} 2 & 1 & 0 & 0 & -8 \\ & & 2 & 4 & 8 \\ \hline & 1 & 2 & 4 & \end{array}$$

Add 0 and 4. Multiply 4 by 2. Write 8 under the second -8.

```
2 | 1   0   0  -8
        2   4   8
    1   2   4   0
```

Add −8 and 8. Your remainder is 0.

Answer: $x^2 + 2x + 4$

Remember to start one degree less than the degree of the original problem, which was 3.

Practice Problems

Perform the following divisions.

1. $(3x - 9) \div 3$
2. $(4x^2 - 6x + 8) \div 2$
3. $(12x^3 - 10x^2) \div 2x$
4. $(5x^3 - 10x^2 + 15x) \div 5x$
5. $(16x^3y - 20x^2y^2) \div 4xy$
6. $(12x^3y^4 - 10x^2y^3 + 20xy^5) \div 2xy^3$
7. $(5x^2y - 8xy^2) \div 5xy$
8. $(2x - 3y + 6) \div 2$

Use long division to find the quotients.

9. $(x^2 - 4x - 5) \div (x - 5)$
10. $(x^2 + x - 12) \div (x - 3)$
11. $(x^2 - 2x - 6) \div (x - 1)$
12. $(x^2 + 5x + 7) \div (x + 1)$
13. $(3x^2 - 2x + 5) \div (x + 2)$
14. $(2x^2 + 4x + 7) \div (x + 3)$
15. $(x^3 - x + 2) \div (x - 1)$
16. $(4x^3 + 2x^2 - 5) \div (x + 3)$
17. $(x^3 - 1) \div (x - 1)$
18. $(8x^3 + 27) \div (2x + 3)$
19. $(2x^3 - 3x^2 + x - 4) \div (x - 2)$
20. $(x^3 - 2x^2 - 3x + 4) \div (x + 1)$

21. $(x^3 - 3x^2 - 4x + 8) \div (x^2 - 2x + 2)$
22. $(2x^3 - x^2 + 4x - 5) \div (x^2 + x + 4)$
23. $(3x^4 - 4x^2 + 6) \div (x^2 - 1)$
24. $(6x^3 - 4x + 5) \div (x^2 + 2)$

Use synthetic division to find the quotients.

25. $(x^2 - 4x - 32) \div (x + 4)$
26. $(x^2 - 10x + 24) \div (x - 6)$
27. $(3x^2 - 2x + 1) \div (x - 2)$
28. $(8x^2 - 6x + 5) \div (x + 1)$
29. $(x^3 - 4x^2 - 6x + 1) \div (x + 2)$
30. $(4x^3 - 5x - 3) \div (x - 3)$

Section 6
Solving Rational Equations

Learning Objectives

When you finish your study of this section, you should be able to
- Solve equations involving rational expressions
- Solve proportions

Solving Rational Equations

To solve equations involving rational expressions, we use the following steps:

Step 1: Multiply through by a least common denominator (LCD) to clear the equation of fractions.

Step 2: Solve the remaining equation, using techniques learned in previous chapters.

Step 3: Once you arrive at an answer(s), check to see if it creates an **undefined** expression in the original equation. In other words, look at the domain of the original problem and make sure your solution(s) are part of it. If one or more of your solutions are not part of the domain, do not write them as a solution. (They are called **extraneous solutions**.)

Let's look at some examples:

Example 1

Solve: $\dfrac{-4}{3}x = \dfrac{2}{3}x + 3$.

The LCD is 3 so we multiply each **term** of the equation by 3.

$$3\left(\dfrac{-4x}{3}\right) = 3\left(\dfrac{2x}{3}\right) + 3(3)$$

$$\cancel{3}\left(\dfrac{-4x}{\cancel{3}}\right) = \cancel{3}\left(\dfrac{2x}{\cancel{3}}\right) + 3(3) \qquad \text{Cancel denominators and multiply.}$$

$-4x = 2x + 9$ \qquad Subtract $2x$ from both sides.

$-4x - 2x = 2x - 2x + 9$ \qquad Combine like terms.

$-6x = 9$ \qquad Divide both sides by -6.

233

Chapter 6: Rational Expressions and Equations

$$\frac{-6x}{-6} = \frac{9}{-6}$$

Answer: $x = \frac{-9}{6} = \frac{-3}{2}$ Remember to reduce fractions to lowest terms and check.

Check: The original problem does not have an x in the denominator, so there are no limitations on the domain. Our answer should work.

Example 2

Solve: $\frac{6}{x} + \frac{2}{3} = \frac{5}{4}$.

The LCD is $12x$. Multiply **each** term by $12x$:

$$(12x)\left(\frac{6}{x}\right) + (12x)\left(\frac{2}{3}\right) = (12x)\left(\frac{5}{4}\right)$$

$$(12\cancel{x})\left(\frac{6}{\cancel{x}}\right) + (\cancel{12}^{4}x)\left(\frac{2}{\cancel{3}}\right) = (\cancel{12}^{3}x)\left(\frac{5}{\cancel{4}}\right)$$ Cancel denominators and multiply.

$72 + 8x = 15x$ Subtract $8x$ from both sides.

$72 = 7x$ Divide both sides by 7.

Answer: $\frac{72}{7} = x$

Check: The domain of the first term of the original equation is all real numbers except 0. Since our answer is not 0, it should work.

Example 3

Solve: $\frac{1}{x+1} + 5 = \frac{3}{2}$.

The LCD is $2(x+1)$. Multiply **each** term by $2(x+1)$

$$2(x+1)\left(\frac{1}{x+1}\right) + 2(x+1)(5) = 2(x+1)\left(\frac{3}{2}\right)$$

$2(\cancel{x+1})\left(\dfrac{1}{\cancel{x+1}}\right) + 2(x+1)(5) = \cancel{2}(x+1)\left(\dfrac{3}{\cancel{2}}\right)$ Cancel denominators and multiply.

$2 + 10(x+1) = 3(x+1)$ Use the distributive property.

$2 + 10x + 10 = 3x + 3$ Combine like terms.

$10x + 12 = 3x + 3$ Subtract $3x$ and 12 from both sides.

$10x + 12 - 3x - 12 = 3x + 3 - 3x - 12$ Combine like terms.

$7x = -9$ Divide both sides by 7.

Answer: $x = \dfrac{-9}{7}$

Check: The domain of the first term of the original equation is all real numbers except -1. Since our answer is not -1, our answer should work.

Example 4

Solve: $\dfrac{x-1}{x-5} = \dfrac{4}{x-5}$.

The LCD is $x - 5$. Multiply **each** term by $x - 5$:

$(x-5)\left(\dfrac{x-1}{x-5}\right) = (x-5)\left(\dfrac{4}{x-5}\right)$

$\cancel{(x-5)}\left(\dfrac{x-1}{\cancel{x-5}}\right) = \cancel{(x-5)}\left(\dfrac{4}{\cancel{x-5}}\right)$ Cancel denominators.

$x - 1 = 4$ Add 1 to both sides.

$x = 5$

Check: The domain of the expression on both the right and left side is all real numbers except 5. That means that 5 will not work as a solution to this equation so this equation has no solution. Write no solution as the final answer.

Answer: no solution

Example 5

Solve: $\dfrac{2x}{x-3} - 5 = \dfrac{10}{3}$.

The LCD is $3(x-3)$. Multiply **each** term by $3(x-3)$:

$$3(x-3)\left(\dfrac{2x}{x-3}\right) - 3(x-3)(5) = 3(x-3)\left(\dfrac{10}{3}\right)$$

$3\cancel{(x-3)}\left(\dfrac{2x}{\cancel{x-3}}\right) - 3(x-3)(5) = \cancel{3}(x-3)\left(\dfrac{10}{\cancel{3}}\right)$ Cancel denominators and multiply.

$3(2x) - 15(x-3) = (x-3)(10)$ Use the distributive property.

$6x - 15x + 45 = 10x - 30$ Combine like terms.

$-9x + 45 = 10x - 30$ Subtract $10x$ and 45 from both sides of the equation.

$-19x = -75$ Divide both sides by -19.

Answer: $x = \dfrac{75}{19}$

Check: The domain of the first term in the original equation is all real numbers except 3. Since our answer is not 3, our answer should work.

Example 6

Solve: $\dfrac{15}{x} + \dfrac{9x-7}{x+2} = 9$.

The LCD is $x(x+2)$. Multiply **each** term by $x(x+2)$:

$$x(x+2)\left(\dfrac{15}{x}\right) + x(x+2)\left(\dfrac{9x-7}{x+2}\right) = x(x+2)(9)$$

$\cancel{x}(x+2)\left(\dfrac{15}{\cancel{x}}\right) + x\cancel{(x+2)}\left(\dfrac{9x-7}{\cancel{x+2}}\right) = x(x+2)(9)$ Cancel denominators.

Chapter 6: Rational Expressions and Equations

$(x+2)15 + x(9x-7) = x(x+2)9$ Use the distributive property.

$15x + 30 + 9x^2 - 7x = 9x^2 + 18x$ Combine like terms.

$9x^2 + 8x + 30 = 9x^2 + 18x$ Subtract $9x^2$ from both sides.

$8x + 30 = 18x$ Subtract $8x$ from both sides.

$30 = 10x$ Divide both sides by 10.

Answer: $3 = x$

Check: The domain for this equation is all real numbers except 0 and -2. Since our answer is not 0 or -2, our answer should work.

Example 7

Solve: $x - \dfrac{4}{3x} = \dfrac{-1}{3}$.

The LCD is $3x$. Multiply **each** term by $3x$:

$$3x(x) - 3x\left(\dfrac{4}{3x}\right) = 3x\left(\dfrac{-1}{3}\right)$$

$3x(x) - 3\cancel{x}\left(\dfrac{4}{\cancel{3x}}\right) = \cancel{3}x\left(\dfrac{-1}{\cancel{3}}\right)$ Cancel denominators.

$3x^2 - 4 = -x$ The equation is quadratic, so gather all terms on the left side.

$3x^2 + x - 4 = 0$ Factor.

$(3x + 4)(x - 1) = 0$ Set each factor equal to 0.

$3x + 4 = 0$ or $x - 1 = 0$ Solve each equation.

$3x = -4$ or $x = 1$

Answer: $x = \dfrac{-4}{3}$ or $x = 1$

Check: The domain this time is all real numbers except 0. Since neither of our answers is 0, both answers should work.

Example 8

Solve: $\dfrac{-18}{x^2 - 9} + \dfrac{3}{x - 3} = \dfrac{5}{x + 3}$.

$x^2 - 9 = (x + 3)(x - 3)$

One denominator is $(x + 3)$ and another is $(x - 3)$. Recall that the LCD contains every unique factor. Therefore, the LCD is $(x + 3)(x - 3)$:

$(x + 3)(x - 3)\left(\dfrac{-18}{(x + 3)(x - 3)}\right) + (x + 3)(x - 3)\left(\dfrac{3}{x - 3}\right) = (x + 3)(x - 3)\left(\dfrac{5}{x + 3}\right)$

Cancel denominators.

$\cancel{(x + 3)}\cancel{(x - 3)}\left(\dfrac{-18}{\cancel{(x + 3)}\cancel{(x - 3)}}\right) + (x + 3)\cancel{(x - 3)}\left(\dfrac{3}{\cancel{x - 3}}\right) = \cancel{(x + 3)}(x - 3)\left(\dfrac{5}{\cancel{x + 3}}\right)$

$-18 + (x + 3)3 = (x - 3)5$ Use the distributive property.

$-18 + 3x + 9 = 5x - 15$ Combine like terms.

$-9 + 3x = 5x - 15$ Subtract $3x$ and add 15 from both sides.

$6 = 2x$ Divide both sides by 2.

$3 = x$

Check: The domain this time is all real numbers except 3 and -3. Since we obtained 3 as our answer, this equation has no solution.

Answer: no solution

Example 9

Solve: $x + \dfrac{5x}{x - 2} = \dfrac{10}{x - 2}$.

The LCD is $x - 2$. Multiply **each** term by $x - 2$:

$(x - 2)(x) + (x - 2)\left(\dfrac{5x}{x - 2}\right) = (x - 2)\left(\dfrac{10}{x - 2}\right)$

Cancel denominators.

$$(x-2)(x) + (x-2)\left(\frac{5x}{x-2}\right) = (x-2)\left(\frac{10}{x-2}\right)$$

$x(x-2) + 5x = 10$ Use the distributive property.

$x^2 - 2x + 5x = 10$ Subtract 10 from both sides.

$x^2 + 3x - 10 = 0$ Factor.

$(x+5)(x-2) = 0$ Set each factor equal to 0.

$x + 5 = 0$ or $x - 2 = 0$ Solve each equation.

$x = -5$ or $x = 2$

Check: The domain this time is all real numbers except 2, so 2 is not a solution to this equation, but -5 should work.

Answer: $x = -5$

Solving Proportions

An equation that involves two fractions (or rational expressions) set equal to each other is called a **proportion**. A general example of a proportion is the following:

$$\frac{a}{b} = \frac{c}{d}.$$

To solve a proportion in this form, find the product of *ad* and *bc*, set them equal to each other, and then solve. This technique is called cross multiplying. We usually solve proportions by cross multiplying, though you can solve proportions by using the method for solving rational equations we have just shown you. (For instance, Example 4 was a proportion.) Let's look at some examples.

Example 10

Solve: $\frac{1}{2} = \frac{x}{10}$.

$\frac{1}{2} = \frac{x}{10}$ Cross multiply.

$1(10) = 2(x)$ Multiply.

$10 = 2x$ Divide both sides by 2.

Answer: $5 = x$

Example 11

Solve: $\frac{3}{5} = \frac{7}{x}$.

$3(x) = 5(7)$ Cross multiply.

$3x = 35$ Divide both sides by 3.

Answer: $x = \frac{35}{3}$

Check: The domain is all real numbers except 0, so the answer should work.

Example 12

Solve: $\frac{x}{4} = \frac{16}{x}$.

$x^2 = 64$ Cross multiply.

$x^2 - 64 = 0$ Factor.

$(x + 8)(x - 8) = 0$ Set each factor equal to 0.

$x + 8 = 0$ or $x - 8 = 0$ Solve each equation.

Answers: $x = -8$ or $x = 8$

Check: The domain is all real numbers except 0, so the answer should work.

Example 13

Solve: $\frac{x+2}{x} = \frac{x+1}{x-3}$.

$(x + 2)(x - 3) = x(x + 1)$ Cross multiply.

$x^2 + 2x - 3x - 6 = x^2 + x$ Combine like terms.

$x^2 - x - 6 = x^2 + x$ Subtract x^2 from both sides.

$-x - 6 = x$ Add x to both sides.

$-6 = 2x$ Divide both sides by 2.

Answer: $-3 = x$

Check: The domain is all real numbers except 0 and 3, so the answers should work.

Practice Problems

Solve the following equations for *x*, if possible.

1. $\frac{1}{3}x - \frac{5}{6} = \frac{-5}{4}x$

2. $\frac{1}{3} - \frac{2}{5}x = \frac{1}{10} + x$

3. $\frac{2}{x} - \frac{5}{2} = \frac{5}{4x}$

4. $\frac{5}{3x} + \frac{1}{5} = \frac{3}{x} - 2$

5. $\frac{2}{x-1} + \frac{5}{2} = 4$

6. $\frac{4}{x-2} + 3 = \frac{6}{x-2}$

7. $\frac{3}{x+3} + 4 = \frac{3}{x+3}$

8. $\frac{8}{5} - \frac{2}{x+1} = 4$

9. $\frac{x-2}{x+3} - \frac{4}{3} = \frac{1}{6}$

10. $\frac{x-4}{x+5} = \frac{x}{x+2}$

11. $\frac{5}{x} + \frac{3}{2} = 2x$

12. $x + \frac{7}{3} = \frac{2}{x}$

13. $\frac{3}{4} + \frac{x+1}{x+2} = \frac{x}{4}$

14. $\frac{2}{5} + \frac{3x}{10(x+3)} = \frac{x}{10}$

241

Chapter 6: Rational Expressions and Equations

15. $\dfrac{5}{x^2-4} + \dfrac{2}{x-2} = \dfrac{3}{x+2}$

16. $\dfrac{-5}{x^2-3x-4} + \dfrac{1}{x-4} = \dfrac{5}{x+1}$

Solve the following proportions for x, if possible.

17. $\dfrac{1}{3} = \dfrac{2}{x}$

18. $\dfrac{x}{5} = \dfrac{6}{10}$

19. $\dfrac{x}{4} = \dfrac{4}{x}$

20. $\dfrac{x}{2} = \dfrac{6}{x}$

21. $\dfrac{x+4}{3} = \dfrac{x-1}{2}$

22. $\dfrac{x+2}{5} = \dfrac{x-3}{5}$

23. $\dfrac{x-1}{x+2} = \dfrac{x-3}{x+4}$

24. $\dfrac{x}{x+7} = \dfrac{x-1}{x}$

25. $\dfrac{x+4}{4} = \dfrac{8}{x}$

26. $\dfrac{1}{2x-1} = \dfrac{x}{3}$

Section 7
Applications of Rational Functions

Learning Objectives

When you finish your study of this section, you should be able to
- Solve word problems involving ratios
- Solve work-related word problems

Solving Word Problems Involving Ratios

Word problems involving ratios can always be set up as proportions. Solving these equations will be just like solving proportions in Section 6. Let's look at some examples:

Example 1

A cyclist can travel 33 miles in 2.5 hours. At this rate, how many miles can s/he travel in 4 hours?

There are two measurement units included in this problem: miles and hours. When we set up the proportion, we must keep the same units in the top and bottom of both sides of the equation. Let's set up the information we have:

$$\frac{33 \text{ miles}}{2.5 \text{ hours}} = \frac{x \text{ miles}}{4 \text{ hours}}$$

Once we have determined the placement of the numbers, we can drop the units and just solve the proportion.

$\frac{33}{2.5} = \frac{x}{4}$ Cross multiply.

$2.5x = 132$ Divide both sides by 2.5.

Answer: $x = 52.8$ miles

Example 2

Mike can eat 30 cheeseburgers per week. At this rate, how many will he eat in a year?

The units in this problem are burgers and weeks or years. We will use weeks for this problem. Recall that a year consists of 52 weeks. Let's set up the information we have:

$$\frac{30 \text{ burgers}}{1 \text{ week}} = \frac{x \text{ burgers}}{52 \text{ weeks}}$$

$$\frac{30}{1} = \frac{x}{52}$$ Cross multiply.

Answer: $x = 1,560$ burgers

Note: The authors do not recommend eating 30 cheeseburgers per week.

Solving Work-Related Word Problems

These word problems are often a bit tricky for students. The basic formula for these problems looks like this:

$$\frac{\text{Together work time}}{\text{Alone work time}} + \frac{\text{Together work time}}{\text{Alone work time}} = 1$$

In other words, if you work with someone else and together you two can complete a job in 4 hours, the time would be considered *together* work time. If you could complete the job by yourself in 8 hours, then the time is considered *alone* work time.

Let's look at an example:

Example 3

If John can install a door in 6 hours, and Mike can do the same job in 10 hours, how long will it take if they work together?

We don't know the *together* work time, so we will call it x. John's *alone* work time is 6 hours, and Mike's *alone* work time is 10 hours.

$\dfrac{x}{6} + \dfrac{x}{10} = 1$ The LCD is 30. (You could also use 60.)

$30\left(\dfrac{x}{6}\right) + 30\left(\dfrac{x}{10}\right) = 30(1)$ Multiply **each** term by the LCD.

$\cancel{30}5\left(\dfrac{x}{\cancel{6}}\right) + \cancel{30}3\left(\dfrac{x}{\cancel{10}}\right) = 30(1)$ Cancel terms before multiplying.

$5x + 3x = 30$ Combine like terms.

$8x = 30$ Divide both sides by 8.

Answer: $x = 3.75$ hours

Example 4

Suppose a mother hands over cleaning the house to her children, Bill and Jerry. The first week Bill is able to clean the house in 5 hours by himself (Jerry claimed to be sick). The next week, Jerry joins Bill in cleaning the house and the two of them are able to finish the job in 3 hours. In the third week, Jerry has to clean the house by himself. How long will it take him if he works alone, assuming he is working at the same rate he did the week before?

$\dfrac{3}{5} + \dfrac{3}{x} = 1$ 3 is the *together* work time. x represents Jerry's *alone* work time. The LCD is $5x$.

$5x\left(\dfrac{3}{5}\right) + 5x\left(\dfrac{3}{x}\right) = 5x(1)$ Multiply **each** term by the LCD.

$\cancel{5}x\left(\dfrac{3}{\cancel{5}}\right) + 5\cancel{x}\left(\dfrac{3}{\cancel{x}}\right) = 5x(1)$ Cancel terms before multiplying.

$3x + 15 = 5x$ Subtract $3x$ from both sides.

$15 = 2x$ Divide both sides by 2.

Answer: 7.5 hours $= x$

Practice Problems

Solve the following word problems, using the methods outlined in Examples 1 and 2.

1. Juanita can read, on average, about three books a month. Approximately how many books can she read in a year?
2. If a scale model is to be drawn at the scale of 1 cm for every 5 ft, then how many centimeters will represent a length of 40 ft?
3. For every 3 ft of shelving, one support rod is needed. If someone needed six support rods, then approximately how many feet of shelving was used?
4. Based on the past records of a dealership, about two trucks are sold for every five cars that are sold. If the dealership sold 20 trucks this month, how many cars would it have sold?
5. A police department estimates that for every 30,000 people in a county there needs to be at least 50 police officers. Using this estimate, at least how many police officers would be needed in a county with a population of 500,000?

Solve the following word problems, using the methods outlined in Examples 3 and 4.

6. If Mary can do a job in 4 hours and Mike can do the same job in 4 hours, how long should it take them if they are working together?
7. If John can do a job in 6 hours and Mike can do the same job in 8 hours, how long will it take if they are working together?

8. If Kayla can complete a job in 6 hours working by herself and then complete the same job working with Aly in 4 hours, how long would it take Aly to complete the job working alone?

9. Antwan can take care of a large lawn in 4 hours working alone. If his friend Marcus helps him out, they can finish the job in 2.5 hours. How long will it take Marcus to take care of the lawn if he works alone?

Chapter 7
Radical Expressions and Equations

Assignment Checklist

What You Should Do	Where?			When?	
Read, view the videos, and then complete the online work for Chapter 7, Section 1	📖	💻	MathXL	After completing Chapter 6	
Read, view the videos, and then complete the online work for Section 2	📖	💻	MathXL	After completing Chapter 7, Section 1	
Read, view the videos, and then complete the online work for Section 3	📖	💻	MathXL	After completing Section 2	
Read, view the videos, and then complete the online work for Section 4	📖	💻	MathXL	After completing Section 3	
Read, view the videos, and then complete the online work for Section 5	📖	💻	MathXL	After completing Section 4	
Read, view the videos, and then complete the online work for Section 6	📖	💻	MathXL	After completing Section 5	
Read, view the videos, and then complete the online work for Section 7	📖	💻	MathXL	After completing Section 6	
Take the quiz on Chapter 7			MathXL	After completing Section 7	
Post questions and respond to other students' questions in the Discussion Board		💻		Anytime	
Other assignments:					
Notes:					

SECTION 1
RADICAL EXPRESSIONS

Learning Objectives

When you finish your study of this section, you should be able to evaluate perfect n^{th} roots of
- Numbers
- Variables
- Radicals that contain fractions

Evaluating Perfect N^{th} Roots

The general form of a radical expression is $\sqrt[n]{b} = a$. The $\sqrt{}$ symbol is called the **radical** symbol, n is called the **index**, and b is called the **radicand**. Read this expression "the nth root of b equals a."

PERFECT SQUARE ROOTS

When simplifying square roots, it is understood that the index is two, and that index is normally left out. In other words, you have probably seen the expression $\sqrt{4}$; however, you have probably never seen the expression $\sqrt[2]{4}$. Even though their meanings are the same, we almost always leave out the 2 on the outside.

In order to evaluate a square root, you need to answer the following question: What number do you need to square to obtain the radicand (the number under the radical sign)? For our problem, $\sqrt{4}$, what number squared equals 4? The answer could be 2 or -2, since squaring either of these numbers equals 4. However, it is understood that when we ask for the square root of a number, we are interested in the **positive** square root (which is also called the **principal root**). When a **negative sign precedes a radical**, such as $-\sqrt{4}$, find the square root of 4 (which is 2) and then just attach the negative sign, giving a final answer of -2. Thus, for any square root expression of the form $\sqrt{b}$, the answer will be a positive number. For any square root expression of the form $-\sqrt{b}$, the answer will be a negative number.

PERFECT CUBE ROOTS

All cube roots are of the form $\sqrt[3]{b}$, where the index is 3. To find a cube root, simply ask yourself what number raised to the 3rd power equals b? As an example, consider $\sqrt[3]{64}$. The answer is 4, since $4^3 = 64$. (We don't have to worry about principal roots with odd powers since there is only one number that satisfies the given requirement.)

Perfect Nth Roots

There are many more roots other than square and cube roots. For example, $\sqrt[5]{32}$ and $\sqrt[4]{256}$ are perfect 5th and 4th roots respectively. $\sqrt[5]{32} = 2$, since $2^5 = 32$; $\sqrt[4]{256} = 4$ (principal root only), since $4^4 = 256$.

In General, Using $\sqrt[n]{b} = a$

If n is positive and even, then you are only looking for the principal (positive) root of b. If n is positive and odd, then you are looking for the only root of b. Sometimes the answer will be negative.

Let's look at some more examples to try to clear this up.

Example 1
Simplify: $\sqrt{25}$.

The question to answer is what number squared equals 25?

Since we know that $5^2 = 25$, 5 is the square root of 25.

Answer: 5

Example 2
Simplify: $\sqrt[3]{64}$.

The question to answer is what number cubed equals 64?

Since $4^3 = 64$, 4 is the cube root of 64.

Answer: 4

Example 3
Simplify: $\sqrt[5]{-32}$.

The question to answer is what number to the 5th power equals -32?

Since $(-2)^5 = -32$, the answer is -2.

Answer: -2

Example 4
Simplify: $-\sqrt{49}$.

Since there is a negative in front of the radical, we are looking for the principal root made negative. In other words, this means $-1 \cdot \sqrt{49} = -1 \cdot 7 = -7$.

Answer: -7

Example 5

Simplify: $\sqrt{-49}$

The question to answer is what number squared equals -49? Well, there isn't any number that you can square that will equal a negative number, so your answer is no solution.

Answer: Not a Real Number.

(Notice in Example 4 the negative was outside the radical.)

Example 6

Simplify: $\sqrt[3]{27}$.

Since $3^3 = 27$, the answer is 3.

Answer: 3

Helpful Table of Perfect Squares, Cubes, Fourths, and Fifths	
Squares	1, 4, 9, 16, 25, 36, 49, 64, 81, 100, …
Cubes	1, 8, 27, 64, 125, 216, …
Fourths	1, 16, 81, 256, 625, …
Fifths	1, 32, 243, 1024, …

Finding Perfect N^{th} Roots of Variables

First, in this book, we always assume that the variables represent non-negative numbers. If that is the case, these problems are easy. Just divide the index of the radical into the variable's exponent. Your quotient is the exponent on the variable in the answer.

Example 7

Simplify: $\sqrt{a^4}$.

2 (remember: for a square root, $n = 2$) goes into 4 twice.

Answer: a^2

Check: Does $(a^2)^2 = a^4$? Yes.

Example 8

Simplify: $\sqrt[3]{x^{12}}$.

3 goes into 12 four times.

Answer: x^4

Check: Does $(x^4)^3 = x^{12}$? Yes.

Example 9

Simplify: $\sqrt[4]{y^{20}}$.

4 goes into 20 five times.

Answer: y^5

Example 10

Simplify: $\sqrt[4]{81x^8}$

$\sqrt[4]{81} \cdot \sqrt[4]{x^8}$

For the first part, since $3^4 = 81$, the answer is 3.

For the second part, 4 goes into 8 twice.

$3 \cdot x^2$

Answer: $3x^2$

N^{th} Roots of Fractions

To evaluate the nth root of a fraction, you find the nth root of the numerator and denominator separately and then reduce the fraction to lowest terms, if possible.

Example 11

Simplify: $\sqrt{\dfrac{49}{64}}$

$\dfrac{\sqrt{49}}{\sqrt{64}}$

Answer: $\dfrac{7}{8}$

Example 12

Simplify: $-\sqrt{\dfrac{36}{81}}$

$-\dfrac{\sqrt{36}}{\sqrt{81}} = -\dfrac{6}{9}$

Since the fraction can be reduced to lowest terms, we must reduce it.

Answer: $-\dfrac{2}{3}$

Example 13

Simplify: $\sqrt{\dfrac{a^2}{b^4}}$

$\dfrac{\sqrt{a^2}}{\sqrt{b^4}} = \dfrac{a}{b^2}$ 2 goes into 2 once. 2 goes into 4 twice.

Answer: $\dfrac{a}{b^2}$

Example 14

Simplify $\sqrt[3]{\dfrac{27x^3}{64y^9}}$

$\dfrac{\sqrt[3]{27x^3}}{\sqrt[3]{64y^9}} = \dfrac{\sqrt[3]{27}\sqrt[3]{x^3}}{\sqrt[3]{64}\sqrt[3]{y^9}} = \dfrac{3x}{4y^3}$

Answer: $\dfrac{3x}{4y^3}$

Check: $\dfrac{3x}{4y^3} \cdot \dfrac{3x}{4y^3} \cdot \dfrac{3x}{4y^3} = \dfrac{27x^3}{64y^9}$.   It checks!

Practice Problems

Simplify the following.

1. $\sqrt{49}$
2. $\sqrt[4]{81}$
3. $\sqrt{121}$
4. $\sqrt{-81}$
5. $\sqrt[3]{27}$
6. $\sqrt[3]{-64}$
7. $-\sqrt{100}$
8. $\sqrt[4]{16}$
9. $\sqrt[4]{-16}$
10. $\sqrt[5]{32}$
11. $\sqrt{x^{12}}$
12. $\sqrt[3]{x^{15}}$
13. $\sqrt[4]{y^8}$
14. $\sqrt[5]{x^{20}}$
15. $\sqrt{x^2 y^8}$
16. $\sqrt[3]{x^3 y^6}$
17. $\sqrt{25x^2}$
18. $\sqrt{36x^4}$
19. $\sqrt[3]{27x^{18}}$
20. $\sqrt[4]{16x^{12}y^4}$
21. $\sqrt{\dfrac{16}{49}}$
22. $\sqrt[3]{\dfrac{125}{64}}$
23. $\sqrt[3]{\dfrac{a^6}{b^3}}$

24. $\sqrt[5]{\dfrac{x^{20}}{y^{25}}}$

25. $\sqrt{\dfrac{9x^6}{25y^2}}$

26. $\sqrt[4]{\dfrac{16x^4}{81y^8}}$

Section 2
Rational Exponents

Learning Objectives

When you finish your study of this section, you should be able to
- Simplify expressions involving rational exponents

Rational Exponents

This entire section can be summarized in one sentence:

$$A^{\frac{B}{n}} = (\sqrt[n]{A})^B$$

In other words, we can rewrite a number raised to a fractional power as a radical expression. The numerator is the exponent and the denominator is the index of the radical. Let's look at some common examples:

- $\sqrt{x} = x^{1/2}$

- $\sqrt[3]{x} = x^{1/3}$

- $\sqrt[4]{x} = x^{1/4}$

Now, let's look at some examples that involve numbers with rational exponents.

Example 1
Simplify: $9^{1/2}$.

$9^{1/2}$ translates in words to the square root of 9 raised to the first power, or $\sqrt{9} = 3$.

Answer: 3

Check: If you want to check this answer on a scientific calculator, type in 9 to the power of ($\frac{1}{2}$). Make sure you put parentheses around the $\frac{1}{2}$ (or any fractional power).

Example 2

Simplify: $64^{2/3}$

$\left(\sqrt[3]{64}\right)^2$ Rewrite. The numerator is the exponent, and the denominator is the index.

4^2 The cube root of 64 is 4. Now, square it.

Answer: 16

Example 3

Simplify: $100^{-1/2}$

$\dfrac{1}{100^{1/2}}$ Remember to use the negative power rules from Chapter 5, Section 1.

$\dfrac{1}{\sqrt{100}}$ Rewrite.

Answer: $\dfrac{1}{10}$

Example 4

Simplify: $\dfrac{1}{9^{-3/2}}$.

$9^{3/2}$ Again, use the negative power rule from Chapter 5, Section 1.

$\left(\sqrt{9}\right)^3$ The numerator is the exponent, and the denominator is the index.

3^3 The square root of 9 is 3. Now, cube it.

Answer: 27

Example 5

Simplify: $\left(\dfrac{4}{25}\right)^{-1/2}$.

$\left(\dfrac{25}{4}\right)^{1/2}$ A fraction raised to a negative power flips the fraction over.

$\dfrac{25^{1/2}}{4^{1/2}}$ Top **and** bottom are raised to the $\frac{1}{2}$ power.

$$\frac{\sqrt{25}}{\sqrt{4}}$$

Rewrite.

Answer: $\frac{5}{2}$

Example 6

Simplify: $\left(\frac{100}{49}\right)^{3/2}$.

$$\frac{100^{3/2}}{49^{3/2}}$$

Top **and** bottom are raised to the $\frac{2}{3}$ power.

$$\frac{(\sqrt{100})^3}{(\sqrt{49})^3}$$

Rewrite.

$$\frac{10^3}{7^3}$$

Answer: $\frac{1000}{343}$

Example 7

Simplify: $(-8)^{2/3}$.

$(\sqrt[3]{-8})^2$ Rewrite.

$(-2)^2$ The cube root of -8 is -2.

Answer: 4

Note: The $\frac{2}{3}$ is in parentheses in this problem just to make it easier to read. The parentheses don't mean anything.

Example 8

Simplify: $(-16)^{5/2}$.

$(\sqrt{-16})^5$ We can't take the square root of -16.

Answer: Not a real number

In the following examples, you may leave your answer as a number (or variable) raised to a fractional power. A quick summary of the rules from Chapter 5, Section 1 is repeated here to assist you in doing these problems.

Summary of Rules	
Product Rule	$b^m \cdot b^n = b^{m+n}$
Power of Zero	If b does not equal zero, then $b^0 = 1$
Quotient	$\dfrac{b^m}{b^n} = b^{m-n}$
Negative Exponent	$b^{-n} = \dfrac{1}{b^n}$ and $\dfrac{1}{b^{-n}} = b^n$
Power	$(b^m)^n = b^{mn}$

Example 9

Simplify: $2^{\frac{1}{5}} \cdot 2^{\frac{1}{3}}$.

$2^{\left(\frac{1}{5}+\frac{1}{3}\right)}$ When we multiply the 2s, we add the exponents.

$2^{\left(\frac{3}{15}+\frac{5}{15}\right)}$ Find a common denominator and then add the numerators.

Answer: $2^{\left(\frac{8}{15}\right)}$

Example 10

Simplify: $a^{\frac{1}{5}} \cdot a^{\frac{3}{2}}$.

$a^{\left(\frac{1}{5}+\frac{3}{2}\right)}$ When we multiply the a's, we add the exponents.

$a^{\left(\frac{2}{10}+\frac{15}{10}\right)}$ Find a common denominator.

Answer: $a^{\left(\frac{17}{10}\right)}$

Example 11

Simplify: $\dfrac{5^{\frac{3}{5}}}{5^{\frac{4}{7}}}$

$5^{\left(\frac{3}{5}-\frac{4}{7}\right)}$

When we divide the 5s, we subtract the exponents.

$5^{\left(\frac{21}{35}-\frac{20}{35}\right)}$

Find a common denominator.

Answer: $5^{\left(\frac{1}{35}\right)}$

Example 12

Simplify: $\dfrac{x^{\frac{2}{3}}}{x^{\frac{3}{4}}}$.

$x^{\left(\frac{2}{3}-\frac{3}{4}\right)}$

When we divide the xs, we subtract the exponents.

$x^{\left(\frac{8}{12}-\frac{9}{12}\right)}$

Find a common denominator.

$x^{\left(\frac{-1}{12}\right)}$

Recall you cannot leave negative exponents in your answer.

Answer: $\dfrac{1}{x^{\left(\frac{1}{12}\right)}}$

According to the definition of negative exponents, we flip the fraction over.

Example 13

Simplify: $\left(2^{\frac{3}{5}}\right)^5$

$2^{\left(\frac{3}{5} \cdot \frac{5}{1}\right)}$

A power is raised to a power, so we multiply the exponents.

2^3

Answer: 8

Example 14

Simplify: $\left(-x^{\frac{2}{3}}\right)^{\frac{-4}{5}}$

$\dfrac{1}{\left(-x^{\frac{2}{3}}\right)^{\frac{4}{5}}}$

According to the definition of a negative exponent, we flip the fraction over.

$\dfrac{1}{\left(-1x^{\frac{2}{3}}\right)^{\frac{4}{5}}}$ The floating negative sign is really -1.

$\dfrac{1}{(-1)^{\frac{4}{5}}\left(x^{\frac{2}{3}}\right)^{\frac{4}{5}}}$ Each factor in the denominator is raised to the $\dfrac{4}{5}$ power.

$(-1)^{\frac{4}{5}} = \left(\sqrt[5]{-1}\right)^4 = (-1)^4 = 1.$

$\dfrac{1}{x^{\left(\frac{2}{3}\cdot\frac{4}{5}\right)}}$ When we raise a power to a power, we multiply exponents.

Answer: $\dfrac{1}{x^{\left(\frac{8}{15}\right)}}$

Practice Problems

Rewrite the following by using radical notation.

1. $5^{\frac{1}{2}}$

2. $4^{\frac{3}{2}}$

3. $x^{\frac{2}{3}}$

4. $y^{\frac{4}{5}}$

5. $6^{\frac{-1}{3}}$

6. $x^{\frac{-3}{4}}$

Simplify the following.

7. $16^{\frac{1}{2}}$

8. $25^{\frac{3}{2}}$

9. $4^{\frac{-1}{2}}$

10. $8^{\frac{-1}{3}}$

11. $125^{\frac{-2}{3}}$

12. $100^{\frac{3}{2}}$

13. $16^{\frac{3}{4}}$

14. $81^{\frac{-1}{4}}$

15. $\dfrac{1}{4^{\frac{-1}{2}}}$

16. $\dfrac{1}{27^{\frac{-2}{3}}}$

17. $\left(\dfrac{9}{16}\right)^{\frac{-1}{2}}$

18. $\left(\dfrac{27}{8}\right)^{\frac{2}{3}}$

19. $\left(\dfrac{16}{81}\right)^{\frac{1}{4}}$

20. $\left(\dfrac{1000}{27}\right)^{\frac{-1}{3}}$

21. $(-16)^{\frac{3}{2}}$

22. $(-27)^{\frac{-1}{3}}$

Simplify the following. You may leave your answer as a number or variable raised to a fractional power.

23. $5^{\frac{1}{2}} \cdot 5^{\frac{1}{3}}$

24. $4^{\frac{2}{3}} \cdot 4^{\frac{3}{5}}$

25. $2^{\frac{-1}{5}} \cdot 2^{\frac{5}{6}}$

26. $7^{\frac{-5}{4}} \cdot 7^{\frac{-3}{2}}$

27. $x^{\frac{1}{4}} \cdot x^{\frac{2}{5}}$

28. $x^{\frac{-8}{5}} \cdot x^{\frac{3}{4}}$

29. $y^0 \cdot y^{\frac{2}{5}}$

30. $a^{\frac{-1}{2}} \cdot a^{\frac{-1}{2}}$

31. $\dfrac{9^{\frac{3}{5}}}{9^{\frac{1}{5}}}$

32. $\dfrac{4^{\frac{1}{3}}}{4^{\frac{2}{3}}}$

33. $\dfrac{6^{\frac{2}{5}}}{6^{\frac{1}{4}}}$

34. $\dfrac{10^{\frac{2}{5}}}{10^{\frac{4}{3}}}$

35. $\dfrac{x^{\frac{-1}{3}}}{x^{\frac{3}{4}}}$

36. $\dfrac{x^{\frac{5}{6}}}{x^{\frac{-1}{3}}}$

37. $\left(5^{\frac{2}{3}}\right)^3$

38. $\left(4^{\frac{1}{5}}\right)^{\frac{5}{7}}$

39. $\left(2^{\frac{-1}{4}}\right)^{\frac{6}{5}}$

40. $\left(-5^{\frac{-5}{2}}\right)^{\frac{-2}{3}}$

41. $\left(x^{\frac{-2}{7}}\right)^{\frac{-7}{2}}$

42. $\left(y^{\frac{8}{9}}\right)^0$

Section 3
Simplifying Radical Expressions

Learning Objectives

When you finish your study of this section, you should be able to
- Simplify radical expressions involving numbers or variables

Simplifying Radical Expressions

Square Roots

In order to simplify a square root, we need to determine if there is a perfect square that divides evenly into the radicand. To help in doing this, we list here the perfect squares less than or equal to 100:

$$4, 9, 16, 25, 36, 49, 64, 81, 100$$

Use this list of perfect squares and the following steps to simplify square roots:

Step 1: Determine if a perfect square divides into the radicand evenly.

Step 2: If a perfect square does divide the radicand, break up the radicand into two radicals multiplied together. The first radical will contain the perfect square; the second will contain the multiple that creates the original radicand. If more than one perfect square evenly divides into the radicand, take the larger of the two.

Step 3: Then, take the square root of the perfect square, and place the answer in front of the other radical that contains the multiple.

If that sounds complicated, you will see that following these rules is actually easier than talking about them!

Example 1
Simplify: $\sqrt{75}$.

Step 1: Find the largest perfect square that divides into 75. The largest perfect square that divides into 75 is 25.

Step 2: Since 25 is the largest, we now rewrite the expression as two radicals.

$$\sqrt{75} = \sqrt{25 \cdot 3} = \sqrt{25} \cdot \sqrt{3}$$

Step 3: Now take the square root of 25, which gives us the following.

Answer: $5\sqrt{3}$

To check for a wrong *answer*: When simplifying radicals, it is important to know that you are changing only the appearance of the radical, not its actual value. Just like simplifying fractions, when you simplify $\frac{6}{12}$ and obtain $\frac{1}{2}$, what you change is only the appearance of the fraction. If you divide 6 by 12, you get .5. This is the same value you obtain when dividing 1 by 2. Now, let's apply this principal to radicals and use a scientific calculator. Your calculator will verify that $\sqrt{75}$ is approximately 8.66. Now, if you multiply 5 by $\sqrt{3}$, you will see that their product is also 8.66. This doesn't **guarantee** that you have the answer completely simplified; however, it does verify that you didn't do anything wrong with your simplifying.

Example 2
Simplify: $\sqrt{28}$.

Step 1: The largest perfect square that divides into 28 is 4.

Step 2: $\sqrt{28} = \sqrt{4 \cdot 7} = \sqrt{4} \cdot \sqrt{7}$

Step 3: Take the square root of 4.

Answer: $2\sqrt{7}$

To check for a wrong *answer:* On a scientific calculator, the original problem and final answer equal approximately 5.292.

Example 3
Simplify: $5\sqrt{48}$.

Step 1: 48 is divisible by 4 and 16, but we use 16, since it is larger.

Step 2: $5\sqrt{48} = 5\sqrt{16 \cdot 3} = 5\sqrt{16} \cdot \sqrt{3}$

Step 3: Now take the square root of 16.

$5 \cdot 4\sqrt{3} = 20\sqrt{3}$

Answer: $20\sqrt{3}$

CUBE ROOTS

In simplifying an expression involving cube roots, we follow the same steps, but we use the following list of perfect cubes instead of the list of perfect squares:

$$8, 27, 64, 125, 216, 343, 512, 729, 1000$$

(A calculator may be helpful.)

Example 4

Simplify: $\sqrt[3]{16}$.

The largest perfect cube that divides into 16 evenly is 8.

Answer: $\sqrt[3]{8} \cdot \sqrt[3]{2} = 2\sqrt[3]{2}$

To check for a wrong *answer*: Again, if you type in the original problem and the final answer into your calculator, you will see that both equal approximately 2.52, as a decimal. If you have trouble doing cube roots on your calculator, be sure to ask your instructor to help you.

Example 5

Simplify: $\sqrt[3]{256}$.

The largest perfect cube that divides into 256 evenly is 64.

$\sqrt[3]{64} \cdot \sqrt[3]{4} = 4\sqrt[3]{4}$

Answer: $4\sqrt[3]{4}$

Example 6

Simplify: $3\sqrt[4]{32}$.

Now, we are looking for the largest perfect fourth power that divides into 32. In this case, it's 2^4, which is 16.

$3\sqrt[4]{16} \cdot \sqrt[4]{2} = 3 \cdot 2\sqrt[4]{2} = 6\sqrt[4]{2}$

Answer: $6\sqrt[4]{2}$

Note: Chapter 7, Section 1 has a table that includes the first few of the perfect squares, cubes, fourths, and fifths. This table can be very helpful in simplifying these problems.

Example 7

Simplify: $\sqrt[5]{96}$.

This time we are looking at perfect fifth powers. In this case, it's 2^5, which is 32.

Answer: $\sqrt[5]{32} \cdot \sqrt[5]{3} = 2\sqrt[5]{3}$

Variables

To simplify a radical expression involving a variable, divide the index into the exponent of the variable. Your quotient is the exponent that goes in the answer outside the radical. Your remainder is the exponent that goes in the answer under the radical.

Some examples should clarify this.

Example 8

Simplify: $\sqrt{x^{11}}$

In a square root, the index is two.

2 divides into 11 five times, with a remainder of 1.

$\sqrt[2]{x^{11}} = x^5 \sqrt{x^1}$

Answer: $x^5\sqrt{x}$

Check: $\left(x^5\sqrt{x}\right)^2 = \left(x^5\right)^2 \cdot \left(\sqrt{x}\right)^2 = x^{10} \cdot x = x^{11}$. It checks!

Example 9

Simplify: $\sqrt{x^8}$.

The index is again two.

2 divides into 8 four times with no remainder.

Answer: x^4

When there is no remainder, nothing is left under the radical. This means the radicand was a perfect square.

Check: $(x^4)^2 = x^8$. It checks!

Example 10

Simplify: $\sqrt{a^{25}}$

2 divides into 25 twelve times, with a remainder of 1.

Answer: $a^{12}\sqrt{a}$

Example 11

Simplify: $\sqrt[3]{x^{14}}$.

3 divides into 14 four times, with a remainder of 2.

Answer: $x^4 \sqrt[3]{x^2}$

Check: $(x^4 \sqrt[3]{x^2})^3 = (x^4)^3 \cdot (\sqrt[3]{x^2})^3 = x^{12} \cdot x^2 = x^{14}$

Example 12

Simplify: $\sqrt[3]{x^{22}}$.

3 divides into 22 seven times, with a remainder of 1.

Answer: $x^7 \sqrt[3]{x}$

Example 13

Simplify: $\sqrt[5]{x^{15} y^{22} z^{40}}$.

It may help to break this up into three parts as follows.

$\sqrt[5]{x^{15}} \cdot \sqrt[5]{y^{22}} \cdot \sqrt[5]{z^{40}}$

5 divides into 15 three times with no remainder. Therefore, no x is left under the radical.

5 divides into 22 four times, with a remainder of 2.

5 divides into 40 eight times, with no remainder. Therefore, no z is left under the radical.

Now merge all the parts into one answer as follows.

Answer: $x^3 y^4 z^8 \sqrt[5]{y^2}$

Example 14

Simplify: $\sqrt{20x^6}$.

Just like in example 13, break this up into two parts before beginning.

$\sqrt{20x^6} = \sqrt{20 \cdot x^6} = \sqrt{20} \cdot \sqrt{x^6}$

$\sqrt{20} = \sqrt{4} \cdot \sqrt{5} = 2\sqrt{5}$ 4 is the largest perfect square in 20.

$\sqrt{x^6} = x^3$ 2 divides into 6 three times, with no remainder.

Now, merge both parts together.

Answer: $2x^3\sqrt{5}$ (Notice the radical always is last in the answer.)

Example 15

Simplify: $5\sqrt[3]{80x^4y^2}$

Break up into two parts: numbers and variables.

$5\sqrt[3]{80} \cdot \sqrt[3]{x^4y^2}$

Simplify the number part first.

$5\sqrt[3]{80} = 5\sqrt[3]{8} \cdot \sqrt[3]{10}$ 8 divides into 80 ten times.

$5 \cdot 2 \cdot \sqrt[3]{10} = 10\sqrt[3]{10}$

Now, simplify the variable part.

$\sqrt[3]{x^4y^2} = x\sqrt[3]{xy^2}$ 3 divides into 4 once, with remainder of 1. 3 does not divide into 2.

Now merge the two parts together:

Answer: $10x\sqrt[3]{10xy^2}$

Practice Problems

Simplify the following.

1. $\sqrt{40}$
2. $\sqrt{27}$
3. $\sqrt{20}$
4. $\sqrt{125}$
5. $3\sqrt{75}$
6. $2\sqrt{98}$
7. $\sqrt[3]{54}$
8. $\sqrt[3]{128}$
9. $5\sqrt[3]{375}$
10. $\sqrt[3]{50}$
11. $\sqrt[4]{48}$
12. $\sqrt[4]{5000}$
13. $\sqrt[5]{64}$
14. $2\sqrt[5]{486}$
15. $\sqrt{-75}$
16. $\sqrt[3]{-24}$
17. $\sqrt[5]{-64}$
18. $\sqrt[4]{-160}$
19. $\sqrt{x^7}$
20. $\sqrt{x^{11}}$
21. $\sqrt[3]{x^4}$
22. $\sqrt[3]{x^5 y^2}$
23. $\sqrt[3]{xy^{10}}$
24. $\sqrt[3]{x^3 y^{13}}$
25. $\sqrt[4]{a^7}$

26. $\sqrt[4]{a^{30}}$

27. $\sqrt[4]{a^{10}b^8}$

28. $\sqrt[4]{a^2 b^8}$

29. $\sqrt[5]{a^{17}}$

30. $\sqrt[5]{a^{21}b^{10}}$

31. $\sqrt{64x^3}$

32. $\sqrt{32x^5}$

33. $4\sqrt{50x^8}$

34. $2\sqrt{12x^2}$

35. $\sqrt[3]{32x^5}$

36. $5\sqrt[3]{125x^6 y^2}$

37. $-2\sqrt[3]{-16x^3 y^5}$

38. $\sqrt[4]{48xy^{14}}$

Section 4
Operations with Radical Expressions

Learning Objectives

When you finish your study of this section, you should be able to
- Add and subtract radical expressions
- Multiply radical expressions

Addition and Subtraction of Radicals

Adding and subtracting radicals follows the same rules as adding and subtracting algebraic expressions. In order to add two radicals together, or subtract one radical from another, they must have the same radicand. If they do, then you simply add/subtract the coefficients. If they do not have the same radicand, then see if one or more of the radicals can be simplified before attempting the operation. Let's look at some examples.

Example 1
Simplify: $3\sqrt{2} + 5\sqrt{2} - 2\sqrt{3}$.

$3\sqrt{2} + 5\sqrt{2} - 2\sqrt{3}$

$(3 + 5)\sqrt{2} - 2\sqrt{3}$

Answer: $8\sqrt{2} - 2\sqrt{3}$

Check: If you want to see if your answer is probably correct, type in the original problem on your calculator to see what the decimal answer is. Then, type in the answer to see if the decimals match. In this problem, both decimals are the same (7.8496), a fact that means we didn't do anything wrong in our simplification.

Example 2
Simplify: $4\sqrt{x} - 5\sqrt{y} + 3\sqrt{xy} - \sqrt{x}$

$(4 - 1)\sqrt{x} - 5\sqrt{y} + 3\sqrt{xy}$

Answer: $3\sqrt{x} - 5\sqrt{y} + 3\sqrt{xy}$

Example 3

Simplify: $\sqrt{3} + \sqrt{48}$

At first glance, it appears that this expression can't be simplified. However, $\sqrt{48}$ can be simplified.

$\sqrt{3} + \sqrt{48}$

$\sqrt{3} + \sqrt{16}\sqrt{3}$

$1\sqrt{3} + 4\sqrt{3}$ Recall that if you don't see a coefficient, it is understood to be 1.

Answer: $5\sqrt{3}$

Check: This is another problem you can verify on a calculator. Both decimal equivalents approximately equal 8.66.

Example 4

Simplify: $6\sqrt{12} + 3\sqrt{3} - 5\sqrt{3}$.

$6\sqrt{12} + 3\sqrt{3} - 5\sqrt{3}$ 4 is the largest perfect square that divides into 12.

$6\sqrt{4} \cdot \sqrt{3} + 3\sqrt{3} - 5\sqrt{3}$ $\sqrt{4} = 2$.

$6 \cdot 2\sqrt{3} + 3\sqrt{3} - 5\sqrt{3}$

$12\sqrt{3} + 3\sqrt{3} - 5\sqrt{3}$ $12 + 3 - 5 = 10$.

Answer: $10\sqrt{3}$

Example 5

Simplify: $4\sqrt[3]{5} - 2\sqrt[3]{40}$.

$4\sqrt[3]{5} - 2\sqrt[3]{40}$ 8 is the largest perfect cube that divides into 40.

$4\sqrt[3]{5} - 2\sqrt[3]{8} \cdot \sqrt[3]{5}$ $\sqrt[3]{8} = 2$.

$4\sqrt[3]{5} - 2 \cdot 2\sqrt[3]{5}$

$4\sqrt[3]{5} - 4\sqrt[3]{5}$

Answer: 0

Would you have guessed the answer to Example 5 was 0 just by looking at the problem?

Multiplying Radical Expressions

For nonnegative real numbers a and b, $\sqrt{a} \cdot \sqrt{b} = \sqrt{ab}$. In other words, the product of two square roots is equal to the square root of the product. Once you have multiplied the radicals together, always check to see if you can simplify the radical(s).

Example 6

Simplify: $\sqrt{3} \cdot \sqrt{5}$.

Answer: $\sqrt{15}$

Example 7

Simplify: $2\sqrt[3]{3} \cdot 3\sqrt[3]{18}$.

As long as the indexes are the **same**, you can multiply the radicals together.

$2\sqrt[3]{3} \cdot 3\sqrt[3]{18}$	Multiply terms on the outside together, and then multiply terms on the inside together.
$6\sqrt[3]{54}$	Simplify the radical.
$6\sqrt[3]{27} \cdot \sqrt[3]{2}$	27 is the largest perfect cube that divides into 27. $\sqrt[3]{27} = 3$
$6 \cdot 3 \cdot \sqrt[3]{2}$	

Answer: $18\sqrt[3]{2}$

Example 8

Simplify: $\sqrt{2x} \cdot \sqrt{12x^3}$

$\sqrt{2x} \cdot \sqrt{12x^3}$	Multiply terms under the radical signs.
$\sqrt{24x^4}$	4 is the largest perfect square that divides into 24.

$\sqrt{4} \cdot \sqrt{6} \cdot \sqrt{x^4}$ $\sqrt{4} = 2$, and for the x, the index 2 divides into 4 two times.

Answer: $2x^2\sqrt{6}$

Example 9
Simplify: $2\sqrt{3} \cdot 4\sqrt{3}$

$8\sqrt{9}$

$8 \cdot 3$

Answer: 24

Example 10
Multiply: $\sqrt{3}(\sqrt{2} + \sqrt{7})$.

$\sqrt{3} \cdot \sqrt{2} + \sqrt{3} \cdot \sqrt{7}$ Use the distributive property.

Answer: $\sqrt{6} + \sqrt{21}$

Since $\sqrt{6}$ and $\sqrt{21}$ are not like terms, they cannot be added together, so this is the final answer.

Example 11
Multiply: $\sqrt{5}(\sqrt{5} + 2)$

$\sqrt{5} \cdot \sqrt{5} + \sqrt{5} \cdot 2$ Use the distributive property.

$\sqrt{25} + 2\sqrt{5}$

Answer: $5 + 2\sqrt{5}$

Example 12
Multiply: $(\sqrt{2} + 3)(\sqrt{2} - 5)$

Using either the distributive property or FOIL results in the following.

$\sqrt{4} - 5\sqrt{2} + 3\sqrt{2} - 15$ Simplify $\sqrt{4}$.

$2 - 5\sqrt{2} + 3\sqrt{2} - 15$ Combine like terms.

Answer: $-13 - 2\sqrt{2}$

Example 13
Multiply: $(2\sqrt{7} - 1)(3\sqrt{2} - 5)$

Using either the distributive property or FOIL results in the following.

$6\sqrt{14} - 10\sqrt{7} - 3\sqrt{2} + 5$

Answer: $6\sqrt{14} - 10\sqrt{7} - 3\sqrt{2} + 5$ Since none of these terms can be simplified, and they are not like terms, this is the final answer.

Practice Problems

Simplify the following:

1. $\sqrt{2} + 5\sqrt{3} - 2\sqrt{2}$
2. $8 + 3\sqrt[3]{4} - \sqrt[3]{4}$
3. $\sqrt{50} + 4\sqrt{8}$
4. $\sqrt{150} - 5\sqrt{24}$
5. $\sqrt[3]{32} + 6\sqrt[3]{4}$
6. $\sqrt[3]{64} + 3\sqrt[3]{24} + \sqrt[3]{243}$
7. $2\sqrt[3]{16} - 5\sqrt[3]{54}$
8. $\sqrt[4]{32} - \sqrt[4]{162}$
9. $3\sqrt{12x} + 4\sqrt{x} - \sqrt{75x}$
10. $5\sqrt{y} - 2\sqrt{16y}$
11. $\sqrt{6} \cdot \sqrt{5}$
12. $\sqrt{2x} \cdot \sqrt{3y}$

13. $\sqrt{3} \cdot \sqrt{6}$

14. $\sqrt[3]{4} \cdot \sqrt[3]{4}$

15. $\sqrt[3]{10} \cdot \sqrt[3]{25}$

16. $\sqrt[4]{2x} \cdot \sqrt[4]{8x^3}$

17. $(6\sqrt{2})(3\sqrt{2})$

18. $\sqrt[3]{5} \cdot 5\sqrt[3]{30}$

19. $3\sqrt{2x} \cdot 4\sqrt{2y}$

20. $\sqrt[4]{9} \cdot \sqrt[4]{27}$

21. $(3\sqrt{2})^2$

22. $(5\sqrt{4x})^2$

23. $(2\sqrt[3]{3})^3$

24. $(-3\sqrt[3]{2x})^3$

25. $\sqrt{2}(\sqrt{2} - \sqrt{6})$

26. $\sqrt{3}(4 - 5\sqrt{3})$

27. $\sqrt[3]{3}(\sqrt[3]{9} + 4\sqrt[3]{2})$

28. $2\sqrt{5}(3\sqrt{5} - 4\sqrt{2})$

29. $3\sqrt[3]{2x}(\sqrt[3]{4x} - 5\sqrt[3]{x^2})$

30. $\sqrt[4]{4}(\sqrt[4]{4} - 3\sqrt[4]{64})$

31. $(3 - \sqrt{5})(2 + 3\sqrt{5})$

32. $(\sqrt{2} - \sqrt{5})(\sqrt{2} - \sqrt{5})$

33. $(\sqrt[3]{5} + 2)(\sqrt[3]{25} - 4)$

34. $(\sqrt[3]{2} + \sqrt[3]{4})(2\sqrt[3]{2} - \sqrt[3]{5})$

35. $(2\sqrt{3} - 5)(2\sqrt{3} + 5)$

36. $(\sqrt{6} + 1)(\sqrt{6} - 1)$

37. $(2\sqrt{2} - \sqrt{3})(3\sqrt{2} + 5\sqrt{3})$

38. $(7\sqrt{5} - 2\sqrt{3})(2\sqrt{5} - \sqrt{3})$

Section 5
Division of Radical Expressions

Learning Objectives

When you finish your study of this section, you should be able to
- Rationalize denominators
- Divide radical expressions
- Use conjugates to rationalize denominators

Division of Radical Expressions

Dividing radicals is a bit more complicated than multiplying them because of the following rule of the game: You are not allowed to leave a radical in the denominator of a fraction. Removing radicals from the denominator of a fraction is called **rationalizing the denominator**. To rationalize the denominator, multiply the numerator and denominator by a fraction consisting of radicals that creates an integer in the denominator. (This fraction must also equal 1 when simplified.) We know that multiplying anything by 1 doesn't change its value; however, it can change how it looks. This is a mathematician's way of performing cosmetic surgery on the original problem. Let's try a few examples and see how it works.

Example 1

Simplify: $\dfrac{3}{\sqrt{5}}$.

$\sqrt{5}$ is in the denominator, so we multiply the numerator and denominator by $\sqrt{5}$, giving us the following:

$\dfrac{3}{\sqrt{5}} \cdot \dfrac{\sqrt{5}}{\sqrt{5}}$ Notice that $\dfrac{\sqrt{5}}{\sqrt{5}} = 1$. This equality means that the actual value of the original expression is not going to change.

$\dfrac{3\sqrt{5}}{\sqrt{25}}$ $\sqrt{25} = 5$

Answer: $\dfrac{3\sqrt{5}}{5}$

Check: If you want to make sure that you haven't done anything wrong, type the original problem into the calculator and find its decimal equivalent. Then, do the same to the answer, and you will see that both of them equal approximately 1.3416, thus verifying that we haven't made a mistake.

Example 2

Simplify: $\sqrt{\dfrac{5}{6}}$.

$\sqrt{6}$ is in the denominator, so we multiply the numerator and denominator by $\sqrt{6}$.

$\dfrac{\sqrt{5}}{\sqrt{6}} \cdot \dfrac{\sqrt{6}}{\sqrt{6}}$ Multiply numerators together and then denominators.

$\dfrac{\sqrt{30}}{\sqrt{36}}$ $\sqrt{36} = 6$

Answer: $\dfrac{\sqrt{30}}{6}$

Check: Again, if you want to make sure that you haven't done anything wrong, type the original problem into your calculator, and then do the same to the final answer. Both will equal approximately 0.91287.

Example 3

Simplify: $\dfrac{2}{\sqrt[3]{3}}$.

This one is a bit trickier. What is the smallest perfect cube that 3 divides into evenly? The answer is 27. To get 27, I would have to multiply 3 by 9:

$\dfrac{2}{\sqrt[3]{3}} \cdot \dfrac{\sqrt[3]{9}}{\sqrt[3]{9}}$ Multiply numerators together and then denominators.

$\dfrac{2\sqrt[3]{9}}{\sqrt[3]{27}}$ $\sqrt[3]{27} = 3$

Answer: $\dfrac{2\sqrt[3]{9}}{3}$

Example 4

Simplify: $\dfrac{x}{\sqrt{y}}$, $y \neq 0$

$\sqrt{y}$ is in the denominator, so we multiply the numerator and denominator by $\sqrt{y}$.

$\dfrac{x}{\sqrt{y}} \cdot \dfrac{\sqrt{y}}{\sqrt{y}}$ Multiply numerators together and then denominators.

$\dfrac{x\sqrt{y}}{\sqrt{y^2}}$ $\sqrt{y^2} = y$.

Answer: $\dfrac{x\sqrt{y}}{y}$

Example 5

Simplify: $\sqrt[3]{\dfrac{4}{x}}$, $x \neq 0$

Here, we need to end up with a power of x that is divisible by 3. The smallest option is 3. To obtain x^3, we need to multiply x by x^2. First separate the radical into two parts before multiplying.

$\dfrac{\sqrt[3]{4}}{\sqrt[3]{x}}$

$\dfrac{\sqrt[3]{4}}{\sqrt[3]{x}} \cdot \dfrac{\sqrt[3]{x^2}}{\sqrt[3]{x^2}}$ Multiply numerators together and then denominators.

$\dfrac{\sqrt[3]{4x^2}}{\sqrt[3]{x^3}}$ $\sqrt[3]{x^3} = x$.

Answer: $\dfrac{\sqrt[3]{4x^2}}{x}$

Example 6

Simplify: $\dfrac{6 + 2\sqrt{3}}{6}$.

Notice that the numerator has a greatest common factor.

$$\frac{6 + 2\sqrt{3}}{6}$$
Factor out the GCF.

$$\frac{\cancel{2}(3 + \sqrt{3})}{\cancel{6}3}$$
Simplify like terms.

Answer: $\dfrac{3 + \sqrt{3}}{3}$

Note: There is no need to rationalize in this problem since there is no radical in the denominator.

Example 7

Simplify: $\dfrac{4 - \sqrt{20}}{2}$.

Notice in the numerator that $\sqrt{20}$ can be simplified. The denominator doesn't contain a radical, so rationalizing is not necessary.

$$\frac{4 - \sqrt{4} \cdot \sqrt{5}}{2}$$
4 is the largest perfect square that divides into 20 evenly.

$$\frac{4 - 2\sqrt{5}}{2}$$
Factor out the GCF.

$$\frac{\cancel{2}(2 - \sqrt{5})}{\cancel{2}}$$
Simplify like terms.

Answer: $2 - \sqrt{5}$

Example 8

Simplify: $\dfrac{5 + \sqrt{32}}{10}$.

$$\frac{5 + \sqrt{32}}{10}$$
$\sqrt{32}$ can be simplified.

$$\frac{5+\sqrt{16}\cdot\sqrt{2}}{10}$$

Answer: $\frac{5+4\sqrt{2}}{10}$

Dividing Radical Expressions

To divide radical expressions, use the following steps:

Step 1: First rewrite the problem as a fraction.

Step 2: Reduce the fraction if possible.

Step 3: If there is a radical remaining in the denominator, rationalize the denominator.

Let's try a few to see this process in action.

Example 9

Simplify: $\sqrt{50} \div \sqrt{5}$

Step 1: Rewrite the problem as a fraction.

$$\frac{\sqrt{50}}{\sqrt{5}}$$

Step 2: Simplify the fraction. In this case, 50 is divisible by 5.

$$\sqrt{\frac{50}{5}}$$

Answer: $\sqrt{10}$

Check: Just as with previous examples in this section, you can verify that you haven't done anything wrong by typing in the original problem into the calculator and then entering the final answer. You will notice that both have the decimal equivalent of 3.1622.

Example 10

Simplify: $\sqrt{5} \div \sqrt{50}$

Step 1: Rewrite as a fraction.

$$\frac{\sqrt{5}}{\sqrt{50}}$$

Chapter 7: Radical Expressions and Equations

$$\sqrt{\frac{5}{50}}$$

Step 2: Reduce to lowest terms.

$$\frac{1}{\sqrt{10}}$$

Step 3: Rationalize by multiplying the numerator and denominator by $\sqrt{10}$.

$$\frac{1}{\sqrt{10}} \cdot \frac{\sqrt{10}}{\sqrt{10}}$$

$$\frac{\sqrt{10}}{\sqrt{100}} \qquad \sqrt{100} = 10$$

Answer: $\dfrac{\sqrt{10}}{10}$

Example 11

Simplify: $\sqrt[3]{6x} \div \sqrt[3]{30x^2}$, $x \neq 0$

Step 1: Rewrite as a fraction.

$$\frac{\sqrt[3]{6x}}{\sqrt[3]{30x^2}}$$

Step 2: Reduce fraction to lowest terms.

$$\frac{1}{\sqrt[3]{5x}}$$

Step 3: Rationalize. We need a perfect cube that 5 divides into evenly; 125 works, so we'll have to multiply 5 by 25. Also, we need to multiply x by x^2 to get x^3.

$$\frac{1}{\sqrt[3]{5x}} \cdot \frac{\sqrt[3]{25x^2}}{\sqrt[3]{25x^2}} \qquad \text{Multiply numerators together and then denominators.}$$

$$\frac{\sqrt[3]{25x^2}}{\sqrt[3]{125x^3}} \qquad \sqrt[3]{125x^3} = 5x$$

Answer: $\dfrac{\sqrt[3]{25x^2}}{5x}$

Rationalizing a Radical Expression by Using Conjugates

The conjugate of a radical expression contains the same terms with the opposite sign between them. For example,

- The conjugate of $\sqrt{x} + 4$ is $\sqrt{x} - 4$.
- The conjugate of $\sqrt{2} - \sqrt{5}$ is $\sqrt{2} + \sqrt{5}$.

What you may find interesting is that when you FOIL conjugates, you always lose the middle term. In other words, using FOIL for $(\sqrt{x} + 4)(\sqrt{x} - 4)$ gives us the following:

$$\sqrt{x^2} - 4\sqrt{x} + 4\sqrt{x} - 16 = x - 16$$

Notice the middle terms cancel out, due to having opposite signs. This is an important concept to understand when using conjugates to simplify a radical expression with a binomial denominator. Let's look at some examples.

Example 12

Simplify: $\dfrac{2}{3 + \sqrt{2}}$

This time a binomial is in the denominator. To deal with the binomial, we need to multiply by the conjugate. The conjugate of $3 + \sqrt{2}$ is $3 - \sqrt{2}$.

$\dfrac{2}{3 + \sqrt{2}} \cdot \dfrac{3 - \sqrt{2}}{3 - \sqrt{2}}$ Multiply numerator and denominator by the conjugate.

$\dfrac{2(3 - \sqrt{2})}{9 + 3\sqrt{2} - 3\sqrt{2} - \sqrt{4}}$ FOIL the denominator.

Answer: $\dfrac{2(3 - \sqrt{2})}{7}$ or $\dfrac{6 - 2\sqrt{2}}{7}$

Example 13

Simplify: $\dfrac{5}{\sqrt{2} - \sqrt{5}}$

The conjugate of the denominator is $\sqrt{2} + \sqrt{5}$.

$\dfrac{5}{\sqrt{2} - \sqrt{5}} \cdot \dfrac{\sqrt{2} + \sqrt{5}}{\sqrt{2} + \sqrt{5}}$ Multiply the numerator and denominator by the conjugate.

$$\frac{5(\sqrt{2}+\sqrt{5})}{\sqrt{4}-\sqrt{25}}$$

$$\frac{5(\sqrt{2}+\sqrt{5})}{-3}$$

Note: It is always a good idea to place the negative sign in the numerator or on the side, rather than in the denominator.

Answer: $\dfrac{-5(\sqrt{2}+\sqrt{5})}{3}$

Practice Problems

Simplify the following.

1. $\dfrac{3}{\sqrt{3}}$

2. $\dfrac{6}{\sqrt{2}}$

3. $\dfrac{3}{\sqrt[3]{2}}$

4. $\dfrac{5}{\sqrt[3]{4}}$

5. $\sqrt{\dfrac{3}{2}}$

6. $\sqrt{\dfrac{10}{12}}$

7. $\sqrt[3]{\dfrac{1}{9}}$

8. $\sqrt[3]{\dfrac{3}{10}}$

9. $\dfrac{5}{\sqrt{x}}$

10. $\sqrt{\dfrac{x}{y^3}}, y \neq 0$

11. $\dfrac{8}{\sqrt[3]{x^2}}, x \neq 0$

12. $\sqrt[3]{\dfrac{6}{5x}}, x \neq 0$

13. $\sqrt{40} \div \sqrt{2}$

14. $\sqrt{30x} \div \sqrt{6x}, x \neq 0$

15. $\sqrt{4} \div \sqrt{12}$

16. $\sqrt{10} \div \sqrt{25x}, x \neq 0$

17. $\sqrt[3]{5} \div \sqrt[3]{4}$

18. $\sqrt[3]{6x} \div \sqrt[3]{5x^2}, x \neq 0$

19. $\sqrt[4]{3} \div \sqrt[4]{2}$

20. $\sqrt[4]{x} \div \sqrt[4]{y^3}, y \neq 0$

21. $\dfrac{3 + 6\sqrt{5}}{3}$

22. $\dfrac{6 - 2\sqrt{7}}{4}$

23. $\dfrac{8 - \sqrt{16}}{2}$

24. $\dfrac{10 + \sqrt{24}}{2}$

25. $\dfrac{2 - \sqrt{18}}{4}$

26. $\dfrac{5 + \sqrt{50}}{15}$

27. $\dfrac{1}{\sqrt{2} - 3}$

28. $\dfrac{4}{2 + \sqrt{5}}$

29. $\dfrac{6}{\sqrt{3}-\sqrt{2}}$

30. $\dfrac{9}{\sqrt{7}+\sqrt{5}}$

Section 6
Solving Equations Involving Radicals and Exponents

Learning Objectives

When you finish your study of this section, you should be able to
- Solve simple equations involving exponents
- Solve equations involving one or two radicals

Solving Equations Involving Exponents

You've learned how to solve linear and quadratic equations. Now, we look at solving equations involving powers other than 1 or 2. Here are the basic rules for solving any equation other than a linear one.

Step 1: Isolate the term with the power by moving all other terms to the other side of the equation.

Step 2: Raise both sides to the reciprocal power. Remember the power rule: when you raise a power to a power, you multiply the exponents. Our goal is to create a power of 1 on the x, which will happen if you multiply by the reciprocal power. One major rule that you must remember: **If you take an even root of both sides of an equation, you must put a $\pm$ sign in front of the number whose root you are taking.**

Step 3: Solve for x using what you've learned in the previous chapters.

Step 4: When it's not too complex, check your solution(s).

If this doesn't make sense now, we think it will after a few examples.

Example 1

Solve: $x^2 = 4$.

You already know one way to solve this equation. Subtract 4 from both sides and then factor and set the factors equal to 0. Here, we introduce a different way to solve this problem. The power on x is 2; the reciprocal of 2 is $\frac{1}{2}$. To raise both sides to the power of $\frac{1}{2}$ is the same as taking the square root of both sides (refer to Section 2). Since the square root is an even root, put a $\pm$ sign in front of the $\sqrt{4}$.

$\sqrt{x^2} = \pm\sqrt{4}$ $\qquad\qquad$ $\sqrt{x^2} = (x^2)^{\frac{1}{2}} = x^1 = x$

$x = \pm\sqrt{4}$

Answer: $x = \pm 2$, or we say the solution set is $\{-2, 2\}$. You can write the solution either way.

Notice that the two solutions are 2 and -2, which are the same answers you would obtain if you factored. Using the $\pm$ sign makes sure that you do not forget that every perfect even root has two values, a positive one and a negative one.

Check: Substitute 2 and -2 into the original equation.

$(2)^2 = 4$

$(-2)^2 = 4$

Example 2

Solve: $x^2 - 10 = 14$

$x^2 - 10 + 10 = 14 + 10$ Add 10 to both sides.

$x^2 = 24$ Take the square root of both sides, which is the same as raising both sides to the $\frac{1}{2}$ power.

$\sqrt{x^2} = \pm\sqrt{24}$

$x = \pm\sqrt{4}\sqrt{6}$ Simplify the radical.

Answer: $x = \pm 2\sqrt{6}$, or the solution set is $\{2\sqrt{6}, -2\sqrt{6}\}$

Check: Be careful checking solutions that contain radicals.

$(2\sqrt{6})^2 - 10 = 14$

$4 \cdot 6 - 10 = 14$

$24 - 10 = 14$

$14 = 14$

(You should verify that $-2\sqrt{6}$ also checks.)

Example 3

Solve: $x^4 + 5 = 37$

$x^4 + 5 - 5 = 37 - 5$ Subtract 5 from both sides.

$x^4 = 32$ Take the fourth root of both sides, which is the same as raising both sides to the $\frac{1}{4}$ power.

$\sqrt[4]{x^4} = \pm\sqrt[4]{32}$

$x = \pm\sqrt[4]{16} \cdot \sqrt[4]{2}$ Simplify the radical.

16 is a perfect 4th root.

Answer: $x = \pm 2\sqrt[4]{2}$ $\sqrt[4]{16} = 2$

Example 4

Solve: $4x^6 = -100$.

$4x^6 = -100$ Divide both sides by 4.

$x^6 = -25$ Raise both sides to the $\frac{1}{6}$ th power, which is the same as taking the sixth root of both sides.

$\sqrt[6]{x^6} = \pm\sqrt[6]{-25}$

Wait! Negative numbers do not have even roots, just odd roots, so this equation has no solution.

Answer: No solution

Example 5

Solve: $(x+3)^2 = 4$.

$\sqrt{(x+3)^2} = \pm\sqrt{4}$ Since the power is 2, take the square root of both sides. Remember to add the ± symbol in front of the 4.

$x + 3 = \pm 2$ Subtract 3 from both sides.

$x = -3 \pm 2$

$x = -3 + 2 = -1$ Write out the two answers.

$x = -3 - 2 = -5$

Answer: $x = -1, -5$, or the solution set is $\{-1, -5\}$

Example 6

Solve: $2(2x-1)^4 = 32$.

$2(2x-1)^4 = 32$	Divide both sides by 2.
$(2x-1)^4 = 16$	Take the fourth root of both sides.
$\sqrt[4]{(2x-1)^4} = \pm\sqrt[4]{16}$	The fourth root of 16 is 2. Remember the $\pm$ sign.
$2x - 1 = \pm 2$	Add 1 to both sides.
$2x = 1 \pm 2$	Divide both sides by 2.
$x = \dfrac{1 \pm 2}{2}$	Write out the two answers.

$x = \dfrac{1+2}{2} = \dfrac{3}{2}$ and $x = \dfrac{1-2}{2} = \dfrac{-1}{2}$

Answer: $x = \dfrac{3}{2}, -\dfrac{1}{2}$, or the solution set is $\left\{\dfrac{3}{2}, -\dfrac{1}{2}\right\}$

Example 7

Solve: $x^3 = 8$.

$\sqrt[3]{x^3} = \sqrt[3]{8}$	Since the power is 3, take the cube root of both sides.

Remember $\sqrt[3]{x} = x^{\frac{1}{3}}$, so we are actually raising both sides to the reciprocal power of 3.

Answer: $x = 2$

When you are taking an odd root, you do not need to use the $\pm$ sign. If you are unsure, substitute -2 into the original equation, and you will see it doesn't work.

Example 8

Solve: $x^5 + 4 = -60$.

$x^5 + 4 - 4 = -60 - 4$	Subtract 4 from both sides.
$x^5 = -64$	Take the fifth root.

$\sqrt[5]{x^5} = \sqrt[5]{-64}$ The largest perfect fifth that divides evenly into -64 is -32.

$x = \sqrt[5]{-32} \cdot \sqrt[5]{2}$ $\sqrt[5]{-32} = -2$

Answer: $x = -2\sqrt[5]{2}$

Again, you do not add the $\pm$ sign since 5 is an odd index.

Example 9

Solve: $(x-4)^3 - 3 = 2$.

$(x-4)^3 = 5$ Add 3 to both sides.

$\sqrt[3]{(x-4)^3} = \sqrt[3]{5}$ Take the cube root.

$x - 4 = \sqrt[3]{5}$ Add 4 to both sides.

Answer: $x = 4 + \sqrt[3]{5}$

Example 10

Solve: $(x-1)^2 = 5$.

$\sqrt{(x-1)^2} = \pm\sqrt{5}$ Take the square root of both sides. Add the $\pm$ sign.

$x - 1 = \pm\sqrt{5}$ Now, add 1 to both sides.

Answer: $x = 1 \pm \sqrt{5}$ Notice that you can't combine the radical and the 1.

Solving Equations Involving Radicals

To solve an equation involving a radical, you raise both sides of the equation to the appropriate reciprocal power after you have **isolated** the radical by itself on one side of the equation. Since radicals are the same as rational exponents, we are really just performing the same steps as we did for Examples 1–10.

Example 11

Solve: $\sqrt{x} = 10$.

The equation involves the square root, so you square both sides of the equation to solve it.

$$(\sqrt{x})^2 = 10^2 \qquad \left((x)^{\frac{1}{2}}\right)^{\frac{2}{1}} = x$$

Answer: $x = 100$

Check: $\sqrt{100} = 10$. Correct!

Example 12

Solve: $\sqrt[3]{x} = -3$.

The equation involves the cube root, so you raise both sides of the equation to the third power.

$$(\sqrt[3]{x})^3 = (-3)^3 \qquad \left((x)^{\frac{1}{3}}\right)^{\frac{3}{1}} = x$$

Answer: $x = -27$

Example 13

Solve: $4\sqrt{x+2} = 4$.

$4\sqrt{x+2} = 4$ Divide both sides by 4.

$\sqrt{x+2} = 1$ The equation involves the square root, so you square both sides of the equation.

$(\sqrt{x+2})^2 = 1^2$

$x + 2 = 1$ Subtract 2 from both sides.

Answer: $x = -1$

Check: $4\sqrt{-1+2} = 4\sqrt{1} = 4$. Correct!

Example 14

Solve: $\sqrt{x+2} = -4$.

$(\sqrt{x+2})^2 = (-4)^2$ Square both sides of the equation.

$x + 2 = 16$ Subtract 2 from both sides.

$x = 14$

Check: $\sqrt{14+2} = \sqrt{16} = 4$. *Not correct.*

This is an example of an **extraneous solution**. This means that we followed the rules correctly, but our answer still does not make the original equation a true sentence, so the original equation has **no solution**.

Answer: No solution

Extraneous roots can only occur in equations involving **even** roots. They won't occur in problems involving **odd** roots.

Note: If you were paying close attention to the original problem, you might think to yourself, "Square roots are always positive or 0. A square root couldn't equal -4." Then, you could have just written down no solution and gone on to the next problem.

Example 15

Solve: $\sqrt[3]{2x+1} - \sqrt[3]{x-2} = 0$

If a problem involves two radicals, arrange the equation so that there is one radical on each side of the equation before raising both sides to the appropriate power.

$\sqrt[3]{2x+1} - \sqrt[3]{x-2} = 0$ Add $\sqrt[3]{x-2}$ to both sides.

$\sqrt[3]{2x+1} = \sqrt[3]{x-2}$ Cube both sides.

$(\sqrt[3]{2x+1})^3 = (\sqrt[3]{x-2})^3$

$2x + 1 = x - 2$ Solve the linear equation.

Answer: $x = -3$

Example 16

Solve: $\sqrt[4]{2x} - 6 = -4$.

$\sqrt[4]{2x} - 6 = -4$ Add 6 to both sides.

$\sqrt[4]{2x} = 2$ Raise both sides to the fourth power.

$(\sqrt[4]{2x})^4 = 2^4$

$2x = 16$ Divide both sides by 2.

Answer: $x = 8$

Check: $\sqrt[4]{2(8)} - 6 = 2 - 6 = -4$. Correct!

Example 17

Solve: $\sqrt{x + 12} = x$.

$(\sqrt{x+12})^2 = (x)^2$ Square both sides.

$x + 12 = x^2$

Since we have an x^2 and an x term, we need to move everything to one side so that we can factor the quadratic equation. This means we will add $-x$ and -12 to both sides.

$0 = x^2 - x - 12$ Factor.

$0 = (x - 4)(x + 3)$ Set factors equal to 0.

$x - 4 = 0; x + 3 = 0$

$x = 4$ and $x = -3$

Check: If $x = 4$, then does $\sqrt{4 + 12} = 4$? Yes.

If $x = -3$, then does $\sqrt{-3 + 12} = -3$? No. It equals 3, so that answer is extraneous.

Answer: 4

Example 18

Solve: $\sqrt{x^2 - 2} = x + 4$.

$(\sqrt{x^2 - 2})^2 = (x + 4)^2$ — Square both sides. FOIL the right side.

$x^2 - 2 = x^2 + 4x + 4x + 16$ — Subtract x^2 from both sides. Combine like terms.

$-2 = 8x + 16$ — Subtract 16 from both sides.

$-18 = 8x$ — Divide both sides by 8.

$\dfrac{-18}{8} = \dfrac{-9}{4} = x$

Answer: $x = \dfrac{-9}{4}$

Check: If $x = \dfrac{-9}{4}$, then we are saying that $\sqrt{\left(\dfrac{-9}{4}\right)^2 - 2} = \dfrac{-9}{4} + 4$.

Let's use a calculator. $\dfrac{-9}{4} = -2.25$.

The left side $= 1.75$. The right side $= -2.25 + 4 = 1.75$. Correct!

Example 19

Rewrite equations involving rational exponents as radicals before trying to solve them. If you don't, you risk missing some of the solutions.

Solve: $x^{\frac{3}{2}} = 8$.

Rewrite: $\sqrt[2]{x^3} = 8$

$(\sqrt{x^3})^2 = 8^2$ — Square both sides to eliminate the radical.

$x^3 = 64$ — Take the cube root of both sides.

$\sqrt[3]{x^3} = \sqrt[3]{64}$

Answer: $x = 4$

Example 20

Solve: $(5x)^{\frac{1}{2}} + 4 = 14$.

Rewrite the equation as $\sqrt{5x} + 4 = 14$. Subtract 4 from both sides.

$(\sqrt{5x})^2 = 10^2$ Square both sides.

$5x = 100$ Divide both sides by 5.

Answer: $x = 20$

Check: $\sqrt{5(20)} + 4 = 14$ Correct!

Example 21

Solve: $(3x - 1)^{\frac{-3}{4}} + 4 = 68$

$(3x - 1)^{\frac{-3}{4}} + 4 = 68$ Subtract 4 from both sides.

$(3x - 1)^{\frac{-3}{4}} = 64$ Rewrite as a radical.

$\dfrac{1}{\sqrt[4]{(3x - 1)^3}} = 64$

$\left(\dfrac{1}{\sqrt[4]{(3x - 1)^3}}\right)^4 = 64^4$ Raise both sides to the fourth power.

$\dfrac{1}{(3x - 1)^3} = 16,777,216$ Simplify.

$\sqrt[3]{\dfrac{1}{(3x - 1)^3}} = \sqrt[3]{16,777,216}$ Take the cube root of both sides.

$\dfrac{1}{3x - 1} = 256$ Cross multiply.

$1 = 256(3x - 1)$ Divide both sides by 256.

$$\frac{1}{256} = 3x - 1 \qquad \text{Add 1 to both sides.}$$

$$\frac{257}{256} = 3x \qquad \text{Divide both sides by 3. (Whew!)}$$

Answer: $\frac{257}{768} = x$

Let's check the answer. $\left(3 \cdot \frac{257}{768} - 1\right)^{-\frac{3}{4}} = \left(\frac{1}{256}\right)^{-\frac{3}{4}} = 256^{\frac{3}{4}} = 4^3 = 64$

Finally, adding 4 gives us 68. Our answer is correct.

Practice Problems

Solve the following equations for x, if possible.

1. $x^2 = 36$
2. $x^2 = 50$
3. $x^2 + 14 = 8$
4. $x^4 = 16$
5. $x^4 - 100 = 62$
6. $x^4 - 10 = -2$
7. $2x^6 = -30$
8. $5x^2 - 4 = 26$
9. $(x - 5)^2 = 9$
10. $(2x - 3)^2 = 16$
11. $(x + 1)^2 - 4 = 10$
12. $(5x + 2)^2 + 7 = 23$
13. $3(x - 6)^2 = 15$
14. $-2(x + 9)^4 = -20$
15. $x^3 = 27$
16. $2x^3 = 108$
17. $-3x^3 = 24$
18. $5x^3 - 10 = 615$

19. $x^5 - 1 = 31$
20. $x^7 + 1 = 0$
21. $(x + 2)^3 - 4 = 4$
22. $(3x - 5)^3 + 9 = 8$
23. $2(x + 3)^3 - 4 = 6$
24. $-3(2x - 7)^3 + 4 = -5$
25. $\sqrt{x} = 4$
26. $\sqrt[3]{2x} = -2$
27. $\sqrt{4x} = -5$
28. $\sqrt[4]{6x} = 1$
29. $\sqrt{x - 8} = 5$
30. $\sqrt{5x - 10} = 2$
31. $\sqrt[3]{x - 6} = 3$
32. $\sqrt[3]{2x - 1} + 1 = 5$
33. $\sqrt[4]{3x - 5} + 6 = 4$
34. $\sqrt[5]{x - 3} + 5 = 6$
35. $5\sqrt{x + 4} = 40$
36. $-2\sqrt[3]{2x + 4} = 8$
37. $\sqrt{x - 2} - \sqrt{x + 3} = 0$
38. $\sqrt[3]{2x + 1} - \sqrt[3]{4x + 2} = 0$
39. $\sqrt{x - 6} = x$
40. $\sqrt{3x - 18} = x$
41. $\sqrt{x^2 + 3} = x - 2$
42. $\sqrt{x^2 - 4} = x + 3$
43. $x^{2/3} = 4$
44. $x^{1/3} = -5$
45. $x^{-3/5} = 27$
46. $2x^{-2/3} = 8$
47. $(3x)^{1/3} - 4 = 2$
48. $(4x)^{1/2} + 1 = 7$
49. $(x + 9)^{-1/2} - 3 = -2$
50. $(2x - 1)^{3/4} + 5 = 69$

Section 7
Complex Numbers

Learning Objectives

When you finish your study of this section, you should be able to
- Define a complex number
- Add, subtract, multiply, and divide complex numbers
- Solve equations whose solutions are complex numbers

Introduction to Complex Numbers

Until now, you have always been told that you can't take the square root of a negative number, because when you square any positive or negative number, the answer is always positive. But, beginning in the 16*th* century (look up the name Bombelli), mathematicians started playing around with the possibility that a negative number could have a square root. Let's now make a definition:

$\sqrt{-1} = i$, so $i^2 = -1$

We define i to be the unique number whose square is -1. Numbers of the form $a + bi$, where a and b are real numbers and $i = \sqrt{-1}$, are called complex numbers. If $b = 0$, then they are just our everyday real numbers. If $a = 0$, they are sometimes called pure imaginary numbers. Here is a chart to assist you in understanding the relationships between different types of numbers:

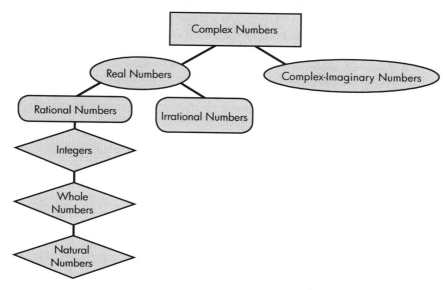

Chapter 7: Radical Expressions and Equations

Operations With Complex Numbers

Adding or subtracting complex numbers is easy; just treat i the same way you would any variable. If a problem involves the multiplication or division of an algebraic expression that contains an i, then you must use the fact that $i^2 = -1$. Also, no answer is considered completely simplified if it contains a square root of a negative number or an i^2 term. Further, just as you weren't allowed to leave a radical in the denominator, you may not leave an i in the denominator. Let's look at some examples.

Example 1

Simplify $(3 + 5i) + (2 - 4i)$.

$(3 + 2) + (5 - 4)i$ Combine like terms.

Answer: $5 + i$

Example 2

Simplify: $(6 - 7i) - (3 - i)$.

$(6 - 7i) - (3 - i)$ Distribute the negative sign.

$6 - 7i - 3 + i$ Combine like terms.

Answer: $3 - 6i$

Example 3

Simplify: $3i(2i - 1)$.

$3i(2i - 1)$ Use the distributive property.

$6i^2 - 3i$ $i^2 = -1$.

Answer: $-6 - 3i$

Note: Recall that you can't leave i^2 in your final answer.

Example 4

Simplify: $(6 + 3i)(4 + 3i)$.

$(6 + 3i)(4 + 3i)$ FOIL

$24 + 18i + 12i + 9i^2$ Combine like terms. $i^2 = -1$.

$24 + 30i - 9$ Combine like terms.

Answer: $15 + 30i$

Example 5
Simplify: i^8.

$(i^2)(i^2)(i^2)(i^2)$ Break this up into i^2 terms.

$(-1)(-1)(-1)(-1)$ Since $i^2 = -1$, each term becomes -1.

Answer: 1

Example 6
Simplify: i^{50}.

$(i^2)^{25}$ Raise a power to a power; multiply exponents.

$(-1)^{25}$ $i^2 = -1$.

Answer: -1

It's worth looking at the pattern of powers of i:

$i^1 = i$

$i^2 = -1$

$i^3 = -i$

$i^4 = 1$

$i^5 = i$ The pattern starts over again.

$i^6 = -1$, etc.

Therefore, one way to raise i to a really large power is to divide that power by 4, and look at the remainder. Find the value in the pattern above and that's the answer. For example, to find i^{3246}, divide 3246 by 4, which is 811 with a remainder of 2, so $i^{3246} = i^2 = -1$. Pretty cool!

Example 7

Simplify: $\dfrac{3}{2i}$

If an i is in the denominator, you must rationalize the fraction because i represents $\sqrt{-1}$, and we can't have radicals in the denominator.

$$\frac{3}{2i} \cdot \frac{i}{i}$$ Multiply the numerator and denominator by i.

$$\frac{3i}{2i^2}$$ $i^2 = -1$.

Answer: $\frac{3i}{-2}$ or $-\frac{3i}{2}$

Example 8

Simplify: $\dfrac{6}{5-i}$

If we have a fraction with a complex number in the bottom, we multiply it by the conjugate the same way we did in Section 5. The conjugate of $a + bi$ is $a - bi$ and vice versa.

$$\frac{6}{5-i} \cdot \frac{5+i}{5+i}$$ Multiply the numerator and denominator by the conjugate.

$$\frac{6(5+i)}{25 - 5i + 5i - i^2}$$ Combine like terms. $i^2 = -1$.

$$\frac{6(5+i)}{25 - (-1)}$$ Combine like terms.

$$\frac{6(5+i)}{26}$$ Divide top and bottom by the GCF, 2.

Answer: $\dfrac{3(5+i)}{13}$

Simplifying Radicals Over the Field of Complex Numbers

At this point, we have solved equations and simplified radicals over the real number system. If we now look at the field of complex numbers (don't worry about what we mean by a "field"), we will be able to solve some equations and simplify some radicals that we couldn't before. Let's look at some examples:

Example 9

Simplify: $\sqrt{-10}$

$\sqrt{-1} \cdot \sqrt{10}$ Break up into -1 times some other number. $\sqrt{-1} = i$.

Answer: $i\sqrt{10}$

Note: To make the expression easier to read, i *comes before a radical but after a number.*

Example 10

Simplify: $5\sqrt{-40}$

$5\sqrt{4}\sqrt{-1}\sqrt{10}$ 4 is the largest perfect square that divides into -40

$5 \cdot 2 \cdot i \cdot \sqrt{10}$ $\sqrt{4} = 2; \sqrt{-1} = i$.

Answer: $10i\sqrt{10}$

Example 11

Simplify: $\sqrt{-3} \cdot \sqrt{-3}$

$i\sqrt{3} \cdot i\sqrt{3}$

$i^2 \sqrt{9}$

$(-1)(3)$

Answer: -3

Note: You can't do this: $\sqrt{-3} \cdot \sqrt{-3} = \sqrt{-3 \cdot -3} = \sqrt{9} = 3$. *This is a common mistake, however. The bottom line is this: when simplifying square roots containing negative numbers, rewrite the problem in terms of i before doing anything else.*

Example 12

Simplify: $\sqrt{2} \cdot \sqrt{-10}$

$\sqrt{2} \cdot \sqrt{-10}$ Rewrite in terms of i.

$\sqrt{2} \cdot i\sqrt{10}$ Multiply the radicals.

$i\sqrt{20}$ $20 = 4 \cdot 5$.

$i\sqrt{4}\sqrt{5}$

Answer: $2i\sqrt{5}$

Example 13

Simplify: $\sqrt{-18} + \sqrt{-50}$.

$\sqrt{-18} + \sqrt{-50}$	Rewrite in terms of i.
$\sqrt{-1}\sqrt{9}\sqrt{2} + \sqrt{-1}\sqrt{25}\sqrt{2}$	$18 = 9 \cdot 2$ and $50 = 25 \cdot 2$
$3i\sqrt{2} + 5i\sqrt{2}$	Combine like terms.

Answer: $8i\sqrt{2}$

Solving Equations Over the Field of Complex Numbers

This process is the same as discussed in the previous section on solving radical equations, except that we will now have solutions for problems containing even roots of negative numbers.

Example 14

Solve: $x^2 = -100$.

$\sqrt{(x^2)} = \pm\sqrt{-100}$	Take the square root of both sides. $\sqrt{-100} = 10i$.

Answer: $x = \pm 10i$

Check: $(10i)^2 = (10i)(10i) = (10)(10)(i)(i) = 100i^2 = (100)(-1) = -100$. Correct!

You should convince yourself that $-10i$ is also a solution.

Example 15

Solve: $4x^2 + 5 = -7$.

$4x^2 + 5 = -7$	Subtract 5 from both sides.
$4x^2 = -12$	Divide both sides by 4.
$x^2 = -3$	Take the square root of both sides.
$\sqrt{(x^2)} = \pm\sqrt{(-3)}$	$\sqrt{-3} = \sqrt{-1} \cdot \sqrt{3}$

Answer: $x = \pm i\sqrt{3}$

Practice Problems

Simplify the following over the field of complex numbers.

1. $(2+3i)+(6-i)$
2. $(4-9i)+(9-4i)$
3. $(1+i)-(3+4i)$
4. $(2-7i)-(5-3i)$
5. $4i(3i-5)$
6. $-5i(2i+4)$
7. $(2+i)(3-2i)$
8. $(4+5i)(3+7i)$
9. $(6-4i)(6+4i)$
10. $(2+3i)(2-3i)$
11. i^4
12. i^{20}
13. i^{30}
14. i^{41}
15. $\dfrac{6}{i}$
16. $\dfrac{7}{2i}$
17. $\dfrac{10}{3i}$
18. $\dfrac{1}{1-i}$
19. $\dfrac{4}{2+3i}$
20. $\dfrac{8}{4-4i}$
21. $\sqrt{-14}$
22. $\sqrt{-40}$
23. $\sqrt{-54}$

24. $4\sqrt{-200}$

25. $\sqrt{-4} \cdot \sqrt{-2}$

26. $\sqrt{-5} \cdot \sqrt{-5}$

27. $3\sqrt{-2} \cdot 5\sqrt{-5}$

28. $4\sqrt{-3} \cdot 7\sqrt{2}$

29. $\sqrt{-4} + \sqrt{-25}$

30. $\sqrt{-8} + \sqrt{-50}$

31. $2\sqrt{-27} - \sqrt{48}$

32. $\sqrt{-20} - 5\sqrt{125}$

Solve for *x* over the field of complex numbers.

33. $x^2 = -64$

34. $x^2 = -72$

35. $4x^2 = -20$

36. $-2x^2 = 64$

37. $3x^2 + 1 = -6$

38. $5x^2 - 2 = 9$

Chapter 8
Quadratic Equations and Inequalities

Assignment Checklist

What You Should Do	Where?			When?	
Read, view the videos, and then complete the online work for Chapter 8, Section 1	📖	💻	MathXL	After completing Chapter 7	
Read, view the videos, and then complete the online work for Section 2	📖	💻	MathXL	After completing Chapter 8 Section 1	
Read, view the videos, and then complete the online work for Section 3	📖	💻	MathXL	After completing Section 2	
Take the quiz on Chapter 8			MathXL	After completing Section 3	
Schedule your final exam with your instructor				After completing the quiz on Chapter 8	
Post questions and respond to other students' questions in the Discussion Board		💻		Anytime	
Other assignments:					
Notes:					

Section 1
Solving Quadratic Equations

Learning Objectives

When you finish your study of this section, you should be able to
- Identify a quadratic equation
- Solve quadratic equations by completing the square
- Solve quadratic equations by using the quadratic formula
- Solving quadratic equations by using the square root property

Solving Quadratic Equations by Using the Square Root Property

This method of solving quadratic equations works **only** on quadratic equations that contain a squared term (or expression). For these types of problems, this method is probably the fastest way to solve them. Here are the steps to follow when using the square root property.

Step 1: Isolate the squared expression on one side of the equation and the number on the other.

Step 2: If there is a coefficient on the squared expression, divide both sides of the equation by the coefficient.

Step 3: Take the square root of both sides of the equation. Place a $\pm$ symbol in front of the number. For example, if you were solving the equation $x^2 = 9$, you would square root both sides at this point. If you don't add a $\pm$ symbol, you would have only one solution, 3; however, if you substitute the solution -3 into the expression, you will see that squaring that solution also equals 9. Therefore, in order to account for both solutions when solving equations like these, you must add the $\pm$ sign at this step of the problem-solving process.

Step 4: Finish solving the equation(s) for x if there are additional values that need to be moved.

Let's look at some examples.

Example 1

Solve $x^2 - 64 = 0$

Step 1: Isolate the squared term on one side by itself by adding 64 to both sides.

$x^2 = 64$

Step 2: Take the square root of both sides. Don't forget the ± sign.

$x = \pm\sqrt{64}$

Step 3: Simplify.

Answer: $x = 8; x = -8$

Check: Substitute each value into the original equation and see if you get zero.

$(8)^2 - 64 = 0$

$64 - 64 = 0$ (True)

$(-8)^2 - 64 = 0$

$64 - 64 = 0$ (True)

Great! Both answers check out.

Example 2

Solve $(3x + 4)^2 = 100$

Step 1: The squared expression is already isolated so nothing needs to be done in this step.

Step 2: Take the square root of both sides.

$3x + 4 = \pm 10$

Step 3: Solve each equation for x.

$3x + 4 = 10$

$3x = 6$

$x = 2$

$3x + 4 = -10$

$3x = -14$

$x = \dfrac{-14}{3}$

Answer: $x = 2$, $x = \dfrac{-14}{3}$

You should check both answers to verify that they make the original equation a true sentence.

Practice Problems

Solve for x using the square-root property.

1. $x^2 = 9$
2. $x^2 = 36$
3. $x^2 - 3 = 10$
4. $x^2 + 5 = -17$
5. $2x^2 - 8 = 16$
6. $3x^2 - 12 = 43$
7. $-x^2 + 5 = 9$
8. $-5x^2 + 2 = 20$
9. $(x + 3)^2 = 16$
10. $(2x - 1)^2 = -25$
11. $(x - 5)^2 = -20$
12. $(3x + 4)^2 = 45$

Solve the following quadratic equations over the field of complex numbers by completing the square.

13. $x^2 + 4x + 6 = 0$
14. $x^2 + 6x + 8 = 0$
15. $x^2 - 2x = 5$
16. $x^2 - 10x = -12$
17. $4x^2 + 16x + 8 = 0$
18. $5x^2 - 20x - 15 = 0$

Solve the following quadratic equations over the field of complex numbers by using the quadratic formula.

19. $x^2 + x + 4 = 0$
20. $x^2 + 2x - 6 = 0$
21. $x^2 - x = 2x + 5$
22. $2x^2 + 5 = x^2 - 3x$
23. $4x^2 + 2x + 5 = 0$
24. $6x^2 - 5x + 4 = 3$
25. $\frac{x^2}{2} + 3x - 1 = 0$
26. $x^2 + \frac{5}{4}x - \frac{1}{8} = 2$

Chapter 8: Quadratic Equations and Inequalities

27. $6x^2 + 3 = 0$
28. $4x^2 - 2x = 0$
29. $-2x^2 + 4x - 6 = 0$
30. $-x^2 - 3x + 7 = 0$

Section 2
Solving Quadratic Inequalities

Learning Objectives

When you finish your study of this section, you should be able to
- Solve quadratic inequalities and write the answers in interval notation

Solving Quadratic Inequalities

Solving a quadratic inequality in a bit more complicated than solving a linear inequality. Once you have solved the quadratic equation, draw a number line, plot the solutions on the number line, and use test points to figure out which sets of numbers make the inequality true. Here are the steps explained for you:

Step 1: If possible, factor the inequality. If it can't be factored, use the quadratic formula.

Step 2: Set each factor equal to zero and solve for x. These values will create the regions you have to test. In other words, if the values are -1 and 6, then you have created three regions: the region below -1, the region between -1 and 6, and the region above 6.

Step 3: Pick a value in each region and substitute it in the original problem to see if it satisfies the original inequality. If it does, then all points in this region will satisfy the original inequality. If not, then no points will satisfy the original equation and thus are not part of the solution set.

Step 4: Use interval notation to write the region(s) of values that satisfies/satisfy the original inequality. Use parentheses for inequalities using the *less than* symbol, $<$, or the *greater than* symbol, $>$, and use brackets for inequalities using the *less than or equal to* symbol, $\leq$, or the *greater than or equal to* symbol, $\geq$.

Let's look at some examples:

Example 1

Solve: $x^2 + 5x + 4 > 0$.

Step 1: Factor the original equation.

If $x^2 + 5x + 4 = 0$, then $(x + 4)(x + 1) = 0$.

Step 2: Set each factor equal to zero and solve for x.

$x + 4 = 0$, and $x + 1 = 0$,

so $x = -4$ and -1.

Step 3: Test each of the three regions by using test points from each region.

The three regions are as follows: numbers less than -4, numbers between -4 and -1, and numbers larger than -1.

Let's choose a test point in each set and see if the test point makes the original inequality a true sentence.

Region 1: Number less than -4

Pick a number less than -4, for example, -5. Is $(-5)^2 + 5(-5) + 4 > 0$? Yes. Numbers in this set work.

Region 2: Numbers between -4 and 1

Pick a number between -4 and -1, for example, -2. Is $(-2)^2 + 5(-2) + 4 > 0$? No. Numbers in this set do not work.

Region 3: Numbers larger than -1

Pick a number larger than -1, for example, 0. (0 is always a good choice; it's easy to substitute in the original equation). Is $0^2 + 5(0) + 4 > 0$? Yes. Numbers in this set work.

Step 4: Write the solution using interval notation. We will use parentheses since the original problem used the *greater than* symbol, $>$.

Thus, the numbers that work are $x < -4$ or $x > -1$.

Answer in interval notation: $(-\infty, -4) \cup (-1, \infty)$

Recall that we combine sets using the union symbol, $\cup$.

Example 2

Solve: $x^2 - x - 6 \leq 0$.

Solve by factoring.

If $x^2 - x - 6 = 0$, then $(x - 3)(x + 2) = 0$, so $x = 3$ and -2.

Draw a number line and plot the points -2 and 3.

Test each of the three regions:

Region 1: Numbers less than -2

Pick a number less than -2, for example, -3. Is $(-3)^2 - (-3) - 6 \leq 0$? No. Numbers in this set do not work.

Region 2: Numbers between -2 and 3

Pick a number between -2 and 3, for example, 0. Is $0^2 - 0 - 6 \leq 0$? Yes. Numbers in this set work.

Region 3: Numbers greater than 3

Pick a number larger than 3, for example 4. Is $4^2 - 4 - 6 \leq 0$? No. Numbers in this set do not work.

Thus, the numbers that work are $-2 \leq x \leq 3$.

Answer in interval notation, $[-2, 3]$

Example 3

Solve: $x^2 + 8x + 16 \geq 0$.

In this case, the problem factors as $(x + 4)^2 = 0$, so $x = -4$. Our number line contains only one number.

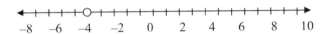

This means that we have only two regions to test: all numbers greater than -4 and all numbers less than -4.

Region 1: All numbers greater than -4

Let's test 0; it's an easy number to substitute.

Is $0^2 + 8(0) + 16 \geq 0$? Yes.

Region 2: All numbers less than -4

Let's test -5.

Is $(-5)^2 + 8(-5) + 16 \geq 0$? Yes.

This means that both regions work, so the final answer is all real numbers.

Answer in interval notation: $(-\infty, \infty)$

Once you realize that the inequality is $(x + 4)^2 \geq 0$, you might reason this way: $(x + 4)^2$ will always be either 0 or larger because when you square any number, you obtain either 0 or a positive number. Thus, you might have then gone directly to all real numbers as the solution.

Caution: If the problem had been $x^2 + 8x + 16 > 0$, all of the numbers less than -4 and greater than -4 would still work, but the number -4 itself wouldn't work since 0 is not greater than 0. In that case, the answer would have been all real numbers except -4.

Example 4

Solve: $x^2 - 7x - 10 \geq 0$.

This inequality doesn't factor, so we will use the quadratic formula. $a = 1$, $b = -7$, and $c = -10$, so

$$x = \frac{-(7) \pm \sqrt{(-7)^2 - 4(1)(-10)}}{2(1)}$$

$$\frac{7 \pm \sqrt{49 + 40}}{2}$$

$$\frac{7 \pm \sqrt{89}}{2}$$

A calculator will verify that the two solutions are around 8.2 and -1.2.

Check the three regions: All numbers less than -1.2, all numbers between -1.2 and 8.2, and all numbers greater than 8.2.

Region 1: All numbers less than -1.2

We will use -2.

$(-2)^2 - 7(-2) - 10 \geq 0$; yes, it checks.

Region 2: All numbers between -1.2 and 8.2

We will use 0.

$(0)^2 - 7(0) - 10 \geq 0$; no, it doesn't check.

Region 3: All numbers greater than 8.2

We will use 10.

$(10)^2 - 7(10) - 10 \geq 0$; yes, it checks.

Since the problem involves a greater than or equal sign, we use brackets, so $x \leq -1.2$ or $x \geq 8.2$.

Answer in interval notation: $\left[-\infty, \frac{7 - \sqrt{89}}{2}\right] \cup \left[\frac{7 + \sqrt{89}}{2}, \infty\right]$

Your instructor may allow you to approximate the solution with decimals, or s/he might want the exact solution shown above. Ask your instructor for her/his preference.

Example 5

Solve: $x^2 - 4x + 4 < 0$

$(x - 2)^2 < 0$.

Now, let's think about this. If we square any number, we obtain either a positive number or 0. In particular, we can't obtain a number less than 0 (unless we are including complex numbers, which we **don't** include when solving quadratic inequalities.)

Answer: There are no values of x that will make this a true sentence, so the inequality has no solution.

Note: If you would have checked the two regions on each side of 2, you would have noticed that neither side would have worked.

Practice Problems

Solve each inequality over the real numbers, and write your answer in interval notation.

1. $x^2 + 5x + 6 \leq 0$
2. $x^2 - 2x - 3 < 0$
3. $x^2 + 10x + 16 > 0$
4. $x^2 - 7x - 18 \geq 0$
5. $3x^2 + 2x - 5 \leq 0$
6. $2x^2 + 7x > -6$
7. $x^2 - 16 \geq 0$
8. $5x^2 - 20x < 0$
9. $x^2 - 6x + 9 \geq 0$
10. $x^2 + 2x + 1 \geq 0$
11. $x^2 + 8x - 16 \leq 0$
12. $x^2 - 4x - 4 \leq 0$
13. $4x^2 - 4x < -1$
14. $9x^2 < -12x - 4$
15. $x^2 + 5x - 7 \leq 0$
16. $2x^2 + x - 4 \leq 0$
17. $x^2 - 3 \geq 0$
18. $x^2 + 3x \geq 5$
19. $x^2 - 7x - 9 < 0$
20. $5x^2 + 2x - 2 > 0$

Section 3
Applications of Quadratic Equations

Learning Objectives

When you finish your study of this section, you should be able to
- Solve problems involving the Pythagorean theorem
- Solve problems by using the height of a falling object formula

The Pythagorean Theorem

Chances are you have heard of the Pythagorean theorem. It's often the only concept people remember from high school geometry. If you don't remember the theorem, here it is:

In a right triangle, with legs a and b and hypotenuse c, $a^2 + b^2 = c^2$.

Let's look at some examples.

Example 1

Solve for x in the right triangle below.

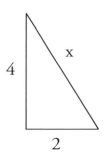

$x^2 = 4^2 + 2^2$	Pythagorean Theorem
$x^2 = 16 + 4$	
$x^2 = 20$	Take the square root of both sides.
$\sqrt{x^2} = \sqrt{20}$	$20 = 4 \cdot 5$.
$x = \sqrt{4} \cdot \sqrt{5}$	
$x = 2\sqrt{5}$, or $x = 4.47$	

Answer: $x = 2\sqrt{5}$, or $x = 4.47$

Note: Notice that we left out the ± symbol in this problem when we took the square root. Why? Well, in this problem, the answer is going to represent the length of the side of a triangle, and lengths can't be negative, so only a positive sign makes sense in this problem.

Example 2

Solve for x in the right triangle below.

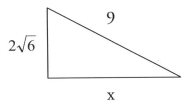

$9^2 = x^2 + \left(2\sqrt{6}\right)^2$ Pythagorean theorem

$81 = x^2 + 24$ Subtract 24 from both sides.

$57 = x^2$ Take the square root of both sides.

$\sqrt{57} = \sqrt{x^2}$

Answer: $\sqrt{57} = x$, or $x = 7.55$

Example 3

A television station needs a 100-ft-tall transmitting tower to send out its broadcasts. The tower must be supported by three cables, each attached to the top of the tower and anchored 50 ft from the base of the tower. How long must each cable be in order to support the tower?

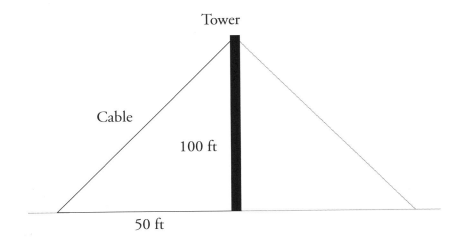

Chapter 8: Quadratic Equations and Inequalities

$$x^2 = 100^2 + 50^2 \qquad \text{Pythagorean theorem}$$

$$x^2 = 10000 + 2500$$

$$x^2 = 12500 \qquad \text{Take the square root of both sides.}$$

$$\sqrt{x^2} = \sqrt{12500}$$

$$x = 111.8$$

Answer: 111.8 ft

Height of a Falling Object Problems

In physics, a common formula for determining the height of an object is the following:

$$S(t) = -16t^2 + v_0 t + s_0,$$

where $S(t)$ is the height in feet after t seconds, v_0 is the initial velocity in feet per second, and s_0 is the initial height in feet.

Let's solve a problem by using this formula.

Example 4

Suppose a baseball is thrown upwards into the air with an initial velocity of 8 feet per second by a person who is 5 feet tall. How long will it take for the baseball to hit the ground?

First of all, we need to determine what variables we have and what we are solving for. We are given $v_0 = 8$, $s_0 = 5$, and $S(t) = 0$ (since the height is 0 when something is on the ground). We are solving for t. Now substitute the values in the formula as follows:

$$0 = -16t^2 + 8t + 5.$$

This quadratic equation doesn't factor, so we will use the quadratic formula to solve for time t.

$a = -16$, $b = 8$, and $c = 5$.

$$\frac{-8 \pm \sqrt{(8)^2 - 4(-16)(5)}}{2(-16)}$$

$$\frac{-8 \pm \sqrt{64 + 320}}{-32}$$

$$\frac{-8 \pm \sqrt{384}}{-32}.$$

Writing these fractions as decimals gives us the following:

$-.36$ and $.86$.

Since a negative answer makes no sense as a period of time, the answer is .86 seconds (which is a bit under one second).

Practice Problems

Find the hypotenuse of the right triangle with the given legs.

1. The legs are 5 and 12.
2. The legs are 7 and 24.
3. The legs are 507 and 676.

Find the missing legs of each right triangle, given a leg and the hypotenuse.

4. The hypotenuse is 50, and the leg is 14.
5. The hypotenuse is $2\sqrt{10}$, and the leg is $2\sqrt{6}$.
6. The hypotenuse is $2\sqrt{13}$, and the leg is 6.

Solve the following word problems.

7. Suppose a baseball is thrown up in the air with an initial velocity of 7 feet per second by a person who is 6.2 feet tall. After how many seconds will it hit the ground?
8. Suppose a baseball is thrown up in the air with an initial velocity of 7.5 feet per second by a person who is 6 feet tall. After how many seconds will it hit the ground?
9. Suppose a baseball is thrown up in the air with an initial velocity of 20 feet per second by Godzilla, who is 246 feet tall. (Don't ask how Godzilla got a baseball in the first place.) After how many seconds will it hit the ground?

Answer Key to Odd Problems

Chapter 1
Section 1

1. Rational
3. Irrational
5. Irrational
7. Rational
9. $\{5,7,9\}$
11. $\{2,3,4,5,6\}$
13. $\{5,6,7\}$
15. B
17. $\{1,2,3,4,5,6,7,9\}$
19. C
21. $\{5,6,7\}$
23. Yes
25. $\{1,2,3,4\}$
27. $\{-4,-3,-2,-1,0,1,2\}$
29. $\emptyset$

Section 2

1. 5
3. -9
5. -1
7. -2
9. -2
11. 3.89
13. 17/28
15. -7
17. 4
19. -2
21. -3
23. 4
25. $-1/10$
27. -1.33
29. 30
31. $-6/7$
33. 11.96
35. -5
37. -106
39. -8
41. Undefined

Section 3

1. 16
3. -64
5. $-8/27$
7. -49
9. 8
11. No real solution
13. 6/5
15. 3
17. 31
19. 5
21. -18
23. -18
25. 2/5
27. -9
29. 10
31. -3
33. 9
35. 3
37. $-15x+20$
39. $6x^2 - 14x$
41. $5x^2 + 2x - 5$
43. $13x - 23$
45. $5x^2 + 10x - 79$

Chapter 2
Section 1

1. 12
3. 2
5. 8
7. 10
9. -8
11. -6
13. $-3/2$
15. 6/25
17. 4
19. -14
21. 4
23. 3
25. 0

27. $-18/13$
29. 0
31. $-11/2$
33. $16/21$
35. $-33/8$
37. $-18/17$ or ≈ 1.06
39. No solution

Section 2

1. 90
3. 45
5. 5
7. $\dfrac{1 \pm \sqrt{33}}{8}$
9. 25.4
11. $\dfrac{4 - y}{3}$
13. $\dfrac{-6 + 5x}{2}$
15. $\dfrac{-9 + x}{2}$
17. $m = F/a$
19. $\dfrac{I}{Pt}$
21. $\dfrac{Y - b}{m}$
23. $\dfrac{V}{\pi r^2}$
25. $\dfrac{3V}{lh}$
27. $\dfrac{2A}{b_1 + b_2}$
29. $-\dfrac{y_2 - y_1}{m} + x_2$
31. $\$8000$
33. $\$5,634.13$
35. $\$22,415.26$

Section 3

1. $n + 4 = 6$
3. $7n = 12$
5. $3x - 8 = 9$
7. $n + 5 = 2n - 5$
9. $7\left(\dfrac{n}{3}\right) = 6$
11. $22, 23$
13. $45, 47$
15. $56, 58, 60$
17. 16 cm
19. $C = 16\pi$ and $A = 64\pi$
21. 7 miles
23. 162 ft^2
25. $\$3000$ at 5%, $\$6000$ at 3%
27. $\$7,350$
29. $\$51,900$

Section 4

1.
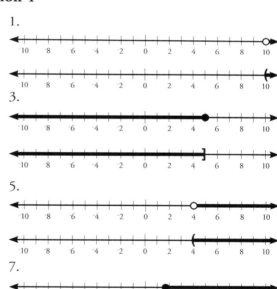

3.

5.

7.

9.
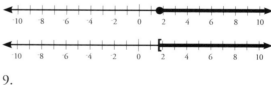

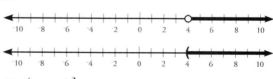

11. $(-\infty, 0\,]$

13. $[3/2, \infty)$
15. $(5, \infty)$
17. $(26/3, \infty)$
19. $[8, \infty)$
21. $(-\infty, \infty)$
23. $(4, 6)$
25. $\emptyset$
27. $[6, \infty)$
29. $(-\infty, \infty)$
31. $(-\infty, -1) \cup (3, \infty)$
33. $(-\infty, -2]$
35. $(-\infty, 3) \cup (8, \infty)$
37. $\emptyset$
39. $[3, 5)$
41.
43.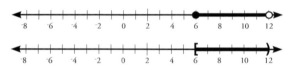
45.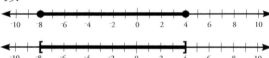

Section 5

1. $x = 2, -2$
3. $x = 14, -4$
5. $x = -3, 9$
7. $x = 7/3, -5/3$
9. $x = 3, 1/3$
11. No solution
13. $x = 1/8$
15. $x = 5, -1$
17. $x = -2, -4$
19. $(-\infty, -3) \cup (3, \infty)$
21. $[-4, 4]$
23. $(-12, 0)$
25. $(-\infty, -9) \cup (3, \infty)$
27. $(1, 13)$
29. $(-\infty, 0) \cup (5, \infty)$
31. $(-\infty, \infty)$
33. $(-15/2, 3/2)$
35. $(-\infty, -7/2] \cup [1/2, \infty)$
37. $(-\infty, 1) \cup (2, \infty)$
39. $2/3$

CHAPTER 3
Section 1

1. 2
3. 1
5. 3
7. No quadrant or x-axis
9. 3
11. -17 odd

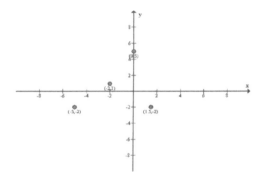

19. $x = 1/2, y = -1$
21. $x = 15, y = -5$
23. $x = 3, y = 2$
25. $x = 9, y = 3$
27. $x = 2, y = -5/3$
29. $x = 0, y = 0$
31. $x = -8, y = 4$

33.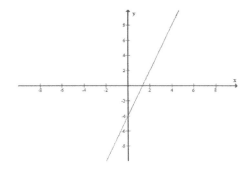

35.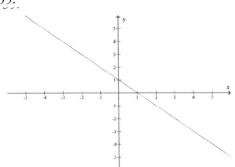

37.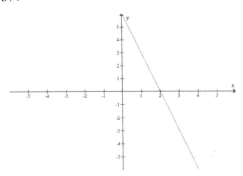

39.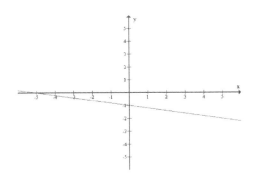

41.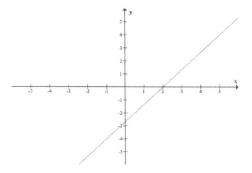

43.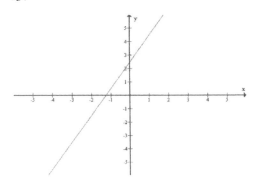

45.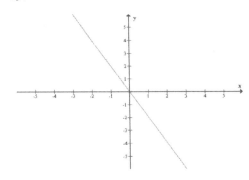

47.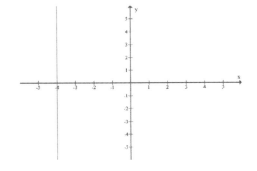

Section 4

1. 6
3. $\dfrac{4x}{5}$
5. $\dfrac{1-x}{x+1}$
7. 2/19
9. 13/24
11. $\dfrac{5x+6}{25-2x}$
13. $\dfrac{4}{12x-5x^2}$
15. $\dfrac{1}{x-y}$
17. $\dfrac{x^2-3x+2}{2x+8}$
19. $\dfrac{(x+5)(3x+14)}{(x+3)(5x+28)}$
21. $\dfrac{3y^2+5y}{3y^2+4y+4}$

Section 5

1. $x-3$
3. $6x^2-5x$
5. $4x^2-5xy$
7. $x-\dfrac{8y}{5}$
9. $x+1$
11. $x-1+\dfrac{-7}{x-1}$
13. $3x-8+\dfrac{21}{x+2}$
15. $x^2+x+\dfrac{2}{x-1}$
17. x^2+x+1
19. $2x^2+x+3+\dfrac{2}{x-2}$
21. $x-1+\dfrac{-8x+10}{x^2-2x+2}$
23. $3x^2-1+\dfrac{5}{x^2-1}$
25. $x-8$
27. $3x+4+\dfrac{9}{x-2}$
29. $x^2-6x+6+\dfrac{-11}{x+2}$

Section 6

1. $x=10/19$
3. $x=3/10$
5. $x=7/3$
7. No solution
9. $x=-13$
11. $x=-5/4, 2$
13. $x=-1$
15. $x=15$
17. $x=6$
19. $x=4,-4$
21. $x=11$
23. $x=-1/2$
25. $x=-8,4$

Section 7

1. 36 books
3. 18 feet
5. $833\tfrac{1}{3}$, so at least 834 (since you can't have $\tfrac{1}{3}$ of a police officer)
7. 24/7 hours
9. 20/3 hours

Chapter 7

Section 1

1. 7
3. 11
5. 3
7. -10
9. Not a real number
11. x^6
13. y^2
15. xy^4
17. $5x$
19. $3x^6$
21. 4/7
23. $\dfrac{a^2}{b}$
25. $\dfrac{3x^3}{5y}$

Section 2
1. $\sqrt{5}$
3. $(\sqrt[3]{x})^2$
5. $\dfrac{1}{\sqrt[3]{6}}$
7. 4
9. $1/2$
11. $1/25$
13. 8
15. 2
17. $4/3$
19. $2/3$
21. Not a real number
23. $5^{\frac{5}{6}}$
25. $2^{\frac{19}{30}}$
27. $x^{\frac{13}{20}}$
29. $y^{\frac{2}{5}}$
31. $9^{\frac{2}{5}}$
33. $6^{\frac{3}{20}}$
35. $\dfrac{1}{x^{\frac{13}{12}}}$
37. 25
39. $\dfrac{1}{2^{\frac{3}{10}}}$
41. x

Section 3
1. $2\sqrt{10}$
3. $2\sqrt{5}$
5. $15\sqrt{3}$
7. $3\sqrt[3]{2}$
9. $25\sqrt[3]{3}$
11. $2\sqrt[4]{3}$
13. $2\sqrt[5]{2}$
15. Not a real number
17. $-2\sqrt[5]{2}$
19. $x^3\sqrt{x}$
21. $x^3\sqrt{x}$
23. $y^3\sqrt[3]{xy}$
25. $a\sqrt[4]{a^3}$
27. $a^2b^2\sqrt[4]{a^2}$
29. $a^3\sqrt[5]{a^2}$
31. $8x\sqrt{x}$
33. $20x^4\sqrt{2}$
35. $2x\sqrt[3]{4x^2}$
37. $4xy\sqrt[3]{2y^2}$

Section 4
1. $5\sqrt{3} - \sqrt{2}$
3. $13\sqrt{2}$
5. $8\sqrt[3]{4}$
7. $-11\sqrt[3]{2}$
9. $4\sqrt{x} + \sqrt{3x}$
11. $\sqrt{30}$
13. $3\sqrt{2}$
15. $5\sqrt[3]{2}$
17. 36
19. $24\sqrt{xy}$
21. 18
23. 24
25. $2 - 2\sqrt{3}$
27. $3 + 4\sqrt[3]{6}$
29. $6\sqrt[3]{x^2} - 15x\sqrt[3]{2}$

31. $-9 + 7\sqrt{5}$
33. $-3 - 4\sqrt[3]{5} + 2\sqrt[3]{25}$
35. -13
37. $-3 + 7\sqrt{6}$

Section 5

1. $\sqrt{3}$
3. $\dfrac{3\sqrt[3]{4}}{2}$
5. $\dfrac{\sqrt{6}}{2}$
7. $\dfrac{\sqrt[3]{81}}{9} = \dfrac{\sqrt[3]{3}}{3}$
9. $\dfrac{5\sqrt{x}}{x}$
11. $\dfrac{8\sqrt[3]{x}}{x}$
13. $2\sqrt{5}$
15. $\dfrac{\sqrt{3}}{3}$
17. $\dfrac{\sqrt[3]{10}}{2}$
19. $\dfrac{\sqrt[4]{24}}{2}$
21. $1 + 2\sqrt{5}$
23. 2
25. $\dfrac{2 - 3\sqrt{2}}{4}$
27. $\dfrac{-(\sqrt{2} + 3)}{7}$
29. $6(\sqrt{3} + \sqrt{2})$

Section 6

1. $x = 6, -6$
3. No real solution
5. $x = 3\sqrt[4]{2}, -3\sqrt[4]{2}$
7. No real solution
9. $x = 8, 2$
11. $x = -1 \pm \sqrt{14}$
13. $x = 6 \pm \sqrt{5}$
15. $x = 3$
17. $x = -2$
19. $x = 2$
21. $x = 0$
23. $x = -3 + \sqrt[3]{5}$
25. $x = 16$
27. No real solution
29. $x = 33$
31. $x = 33$
33. No real solution
35. $x = 60$
37. No real solution
39. No real solution
41. No real solution
43. $x = 8, -8$
45. $x = 1/243$
47. $x = 72$
49. $x = -8$

Section 7

1. $8 + 2i$
3. $-2 - 3i$
5. $-12 - 20i$
7. $8 - i$
9. 52
11. 1
13. -1
15. $-6i$
17. $\dfrac{-10i}{3}$
19. $\dfrac{8 - 12i}{13}$
21. $i\sqrt{14}$
23. $3i\sqrt{6}$

25. $-2\sqrt{2}$

27. $-15\sqrt{10}$

29. $7i$

31. $(6i-4)\sqrt{3}$

33. $x = 8i, -8i$

35. $x = \pm i\sqrt{5}$

37. $x = \dfrac{\pm i\sqrt{21}}{3}$

CHAPTER 8
Section 1

1. $x = 3 \text{ or } -3$

3. $x = \pm\sqrt{13}$

5. $x = \pm 2\sqrt{3}$

7. $x = \pm 2i$

9. $x = 1 \text{ or } -7$

11. $x = 5 \pm 2i\sqrt{5}$

13. $x = -2 \pm i\sqrt{2}$

15. $x = 1 \pm \sqrt{6}$

17. $x = -2 \pm \sqrt{2}$

19. $x = \dfrac{-1 \pm i\sqrt{15}}{2}$

21. $x = \dfrac{3 \pm \sqrt{29}}{2}$

23. $x = \dfrac{-1 \pm i\sqrt{19}}{4}$

25. $x = -3 \pm \sqrt{11}$

27. $x = \dfrac{\pm i\sqrt{2}}{2}$

29. $x = 1 \pm i\sqrt{2}$

Section 2

1. $[-3, -2]$

3. $(-\infty, -8) \cup (-2, \infty)$

5. $[-5/3, 1]$

7. $(-\infty, -4) \cup (4, \infty)$

9. $(-\infty, \infty)$

11. $[-4 - 4\sqrt{2}, -4 + 4\sqrt{2}]$

13. No solution

15. $\left[\dfrac{-5 - \sqrt{53}}{2}, \dfrac{-5 + \sqrt{53}}{2}\right]$

17. $(-\infty, -\sqrt{3}] \cup [\sqrt{3}, \infty)$

19. $\left(\dfrac{7 - \sqrt{85}}{2}, \dfrac{7 + \sqrt{85}}{2}\right)$

Section 3

1. 13

3. 845

5. 4

7. $t = 0.88$ seconds

9. $t = 4.60$ seconds

Course Authors

JERRY SHAWVER
Co-Lead Developer

Jerry Shawver received a B.A. degree in mathematics and education from the University of North Florida and a master's degree in integrated learning technologies from Jacksonville University. He currently teaches classes ranging from college math to statistics and algebra, as well as education and educational technology, at Florida State College at Jacksonville. Shawver taught high school for 11 years before becoming a tenured faculty member at Florida State College at Jacksonville. He is certified to teach this course.

BILL MEISEL
Co-Lead Developer

Bill Meisel received both a B.S. and a B.A. from Florida State University and an M.S. degree in mathematics from the University of North Florida. He currently teaches mathematics, statistics, and calculus classes at Florida State College at Jacksonville, where he also co-chaired the college prep council for 2 years. He is certified to teach this course.

Index

Absolute Value 8
    Equations 62-63
    Functions 107
    Graphs 110
Addition of
    Polynomials 161
    Radicals 272
    Rational expressions 207-213
    Real numbers 9-10
Area of
    Circle 45
    Parallelogram 44
    Rectangle 44
    Square 44
    Trapezoid 45
    Triangle 44
Base (of exponents) 15
Circle 45
    Area 45
    Circumference 45
Commission Problems 47
Completing the Square 313-315
Complex Fractions 216
    Simplifying 216
Complex Numbers 300
    Complex imaginary 300
    Operations with 3001
Compound Interest 37, 156
Conjugates 284
Consecutive Numbers 42
Constants 29
Contradiction 32, 53
Coordinates 71
Cube Roots 266
Degree of Polynomial 160

Dependent Variable 108
Difference of Two Squares 170
Distributive Property 20
Division of Polynomials 223
    Radical expressions 278
    Rational expressions 202
Synthetic 229
Domain of Functions 107
    Rational functions 194
Elements 5
Elimination Method 125
Empty Set 5
Evaluate
    Algebraic expressions 18
    Formulas 35
    Nth roots 249
    Rational functions 193
Exponent 15, 147, 256
    Definition 15
    Equations 288
    Negative 151
    Rules 148
    Zero 149
Exponent Rules 148
    Power 152
    Product 148
    Quotient 150
Extraneous Solutions 233, 294
Factoring Methods 167
    Difference of squares 170
    GCF 167
    General strategy 183
    Grouping 169
    Multi-step 182
    Substitution 181

- Sum/Difference of cubes 171
- Trial and Error 180
- Trinomials 175
- Falling Object 328
- FOIL 164
- Formulas
  - Compound interest 37, 156
  - Simple interest 37, 46
- Functions 107
  - Absolute value 110
  - Domain of 107
  - Linear 110
  - Quadratic 110
  - Range of 107
  - Square root 110
  - Types of 110
- GCF 167
- Graphing
  - Absolute value 112
  - Horizontal lines 77
  - Intercept method 74
  - Large scales 78
  - Linear equations 72
  - Linear functions 72, 110
  - Linear inequalities 50, 102
  - Quadratic equations 113
  - Square root equations 111
  - Systems of inequalities 138
- Grouping Method 169
- Horizontal Lines 77
- Identity 32, 53
- Independent Variable 108
- Index 249, 267
- Indicator Sign 175
- Inequalities
  - Compound 54
  - Linear 50
  - Solving 50
- Integers 3, 300
- Intercept Method 74
- Intersection of Sets 6, 54
- Interval Notation 52
- Irrational Numbers 4, 300
- LCD 209, 217
- Like Terms 21, 162
- Linear Equations
  - Graphing 72
  - Solving 27
- Linear Functions 72, 110
- Multiplication
  - In Scientific notation 155
  - Of Polynomials 163
  - Of Radicals 274
  - Of Real numbers 11
- Nth Roots 249
  - Evaluation of 249
  - Of fractions 252
  - Of variables 251
- Number Line Graphs 322
- Numbers (sets of)
  - Complex 300
  - Integers 300
  - Irrational 300
  - Natural 300
  - Rational 300
  - Real 300
  - Whole 300
- Order of Operations 17
- Parabola 113
- Parallel Lines 85
- Parallelogram 44, 86
- Perfect
  - Cube roots 249
  - Cubes 171, 251
  - Fifths 251
  - Fourths 251

Nth roots	251
Square roots	264
Squares	251
Perimeter of	43
Rectangle	44
Perpendicular Lines	85
Point-Slope Formula	94
Polynomials	160
Addition of	161
Degree of	160
Division of	224
Factoring of	167-173
FOIL	164
Multiplication of	163
Subtraction of	162
Types of	160
Power Rule	152
Principal Root	249
Principle of Zero Product	185
Product Rule	148
Pythagorean Theorem	326
Quadrants	71
Quadratic Equations	185, 311
Graphing	110
Solving	311-318
Quadratic Formula	316-318
Quotient Rule	150
Radical Expressions	
Add/Subtract	272
Division of	278
Multiplication of	274
Rationalizing	278-285
Simplifying	264
Radicals	249, 264
Radicand	249
Range	107
Rational Expressions	193
Add/Subtract	207
Division of	202
Domain of	194
Evaluating	193
Multiplication of	200
Simplification of	195
Rational Functions	193
Rational Numbers	3, 4, 300
Ratios	243
Real Numbers	3, 4, 300
Addition of	9-10
Division/Multiplication of	11-13
Subtraction of	10-11
Rectangles	44
Rectangular Coordinate System	71
Relations	107
Right Triangles	86
Roster Notation	5
Scientific Notation	154
Division	155
Multiplication	155
Set Notation	5
Sets	
Intersection of	6, 54
Union of	6, 54
Simple Interest Formula	37, 46
Slope	81
Formula	81
From equation	84
From graph	82
Of parallel lines	85
Of perpendicular lines	85
Slope-Intercept Formula	92
Solving	
Absolute value equations	62
Exponential equations	288
Inequalities	50
Linear equations	27-32
Proportions	239
Quadratic equations	185, 311-318
Quadratic inequalities	321

    Radical equations 292

    Rational equations................... 233

    Ratios 243

    Systems of equations 121-130

    Systems of inequalities.............. 138

    Work problems........................ 244

Square Root 16, 264

Square Root Function 110

Square Root Property 311

Standard Form 92

Subsets 5

Synthetic Division.......................... 229

Systems of Equations............... 121-130

Systems of Inequalities 138

Trial and Error Method 180

Undefined 233

Union of Sets 6, 54

Variables

    Dependent 108

    Independent............................ 108

Vertical Line Test........................... 108

Vertical Lines................................... 77

Whole Numbers....................... 3, 300

Writing Equations of Lines.......... 92-98